598.91 P822 11076

Poole, Alan Forsyth.

Ospreys

APR 28 '94

APR 04 1995

Apr 17 1995

Harrisburg Area Community College
McCormick Library
3300 Cameron Street Road
Harrisburg, PA 17110

OSPREYS:
A Natural and
Unnatural History

OSPREYS
A NATURAL AND UNNATURAL HISTORY

ALAN F. POOLE
Manomet Bird Observatory,
Manomet, Massachusetts

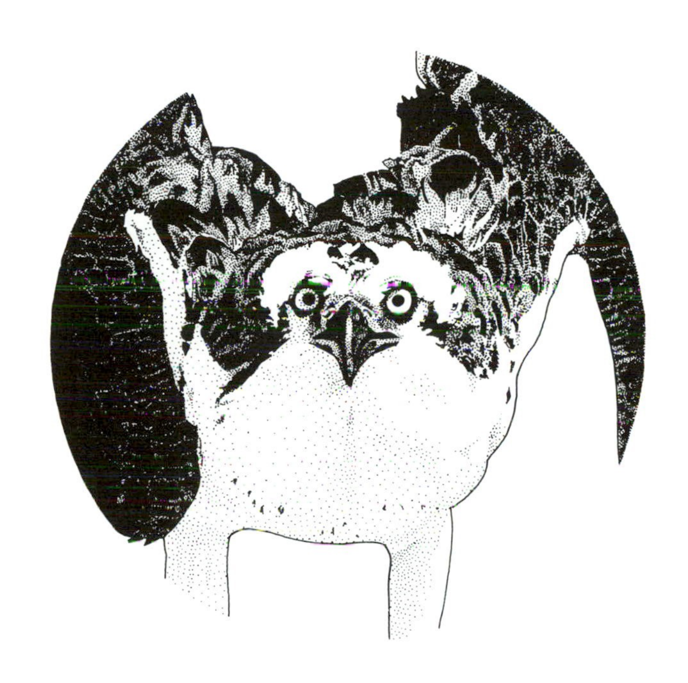

CAMBRIDGE UNIVERSITY PRESS
Cambridge
New York   New Rochelle   Melbourne   Sydney

Published by the Press Syndicate of the University of Cambridge
The Pitt Building, Trumpington Street, Cambridge CB2 1RP
32 East 57th Street, New York, NY 10022, USA
10 Stamford Road, Oakleigh, Melbourne 3166, Australia

© Cambridge University Press 1989
Line illustrations © Margaret LaFarge

First published 1989

Printed in Great Britain by Scotprint Musselburgh

British Library cataloguing in publication data

Poole, Alan F.
Ospreys
1. Ospreys
I. Title
598′.917

ISBN 0 521 30623 X

VN

For Phoebe Akin Poole, a nestling still . . . And with the memory of highblue summer days on Osprey marshes.

117413

Contents

Foreword

During the 1930s and 1940s when I did so much birding along the shores of New Jersey, Long Island and New England, the Osprey was in very good shape. It symbolized the New England coast more than any other bird. Indeed, I incorporated it into the design for the logo of Camp Chewonki in Maine where I spent so many summers.

Describing a boat trip in Maine in *Birds Over America* (1948), I wrote:

> Ospreys lift on heavily-beating wings from their bulky nests, and protest our passing with an annoyed *chewk, chewk, chewk, chewk*. Since the early 1900s, these white-bellied 'fish hawks' seem to have increased here like all the other fish-eating birds.

That was written just before the widespread use of DDT. .

When James Fisher and I made our grand tour around the perimeter of 'Wild America' in 1953, my British colleague was very much impressed by the number of Osprey eyries we saw in Rhode Island. It was mid-April and these spectacular raptors had recently returned from the south. We saw several of their huge nests, including one on a cartwheel on a pole. Many people living near the water had erected these platforms and Robert Cushman Murphy, whom we visited the next day on Long Island, had a new one at the bottom of the garden with nothing on it but hopes. The hopes were good ones, for along that part of the coast the Osprey was a common, everyday bird. Like the Stork in Europe it was protected by public opinion.

In Britain, 100 years had passed since Ospreys were familiar nesting birds. The Highland lochs in Scotland were their last refuge, from which they were systematically exterminated by collectors. I remember a discussion that James Fisher, Peter Scott and I had over tea with Lord Hurcomb in 1950 when we were in Sweden where

Ospreys still flourished. Couldn't Swedish Osprey eggs be transported to Scotland? There, on the shore of some likely loch they could be hatched and hand-reared. If there was anything to 'imprinting' – the theory that young birds are conditioned by their surroundings during their first few hours or days of life – the Osprey might become re-established in the British Isles.

Today the Osprey is again a local resident of the Scottish lochs, a case history of restoration that is given in detail by Dr Poole in Chapter 11 of this book.

When I moved to Connecticut in 1954 there were approximately 150 occupied Osprey nests within a 10-mile radius of our new home in Old Lyme. There was even an abandoned nest in a dead oak on the ridge above the house. The following season when I surveyed the concentration of nests on Great Island near the mouth of the Connecticut River, I noticed that very few young were to be seen in any of the nests. It was the first week of July and almost no young had been fledged.

Alarmed by this, I remembered that during the previous several years, my friend Charles Broley had noticed that his Florida Bald Eagles were in trouble. Year by year they produced fewer young. He suspected it was because of DDT ingested with the dead fish that had been killed by this residual biocide. It was only a hunch, but it proved to be prophetic. Unfortunately, soon after, Charles Broley died of a heart attack while fighting a brush fire near his summer home in Canada.

I wondered, was DDT also affecting our local Ospreys? Because their diet was exclusively fish, they would be even more vulnerable than Bald Eagles. At my suggestion two biology students at Yale – Peter Ames and Tom Lovejoy – took up the investigation as part of their graduate studies. So did Paul Spitzer, who eventually got his degree at Cornell. There proved to be no question that the population was declining and that DDT was involved. Later I took part in the hearings held by Senator Ribicoff at the state capital to have the chemical banned. Remember, if you will, much of our Osprey work in Connecticut was done even before Rachel Carson published her landmark book, *Silent Spring*, in 1962.

At its low point our local population of Ospreys dipped to a mere nine pairs, but now that DDT has been eliminated from the wetland environment they are recovering and it is but a matter of time before their population may again reach full strength. There could eventually be even more Ospreys than before because so many people are now putting up platforms. Whereas in the old days as many as 50 nests were removed each year from transformer boxes and utility poles in Connecticut, the state now erects an equal number of platforms to encourage Ospreys to switch sites. They are no longer regarded as 'trash' birds.

Alan Poole of Woods Hole has taken New England's favorite bird and put it in global perspective, giving us case histories of its decline and recovery as well as chapters on management, breeding biology and a view toward the future. Margaret LaFarge's drawings add a pleasing element to Dr Poole's well-researched text.

Of all the raptors, the Osprey is the one that can live most happily with modern man, if given a chance.

Roger Tory Peterson

Preface

Senegalese fishermen sing of Ospreys as they paddle dugout canoes through Atlantic surf to tend their nets. New England clammers, normally reticent men, wax garrulous when Ospreys are mentioned. The value of a Swedish lakeside home is said to escalate if Ospreys nest nearby. And several Scottish bus tours now include an Osprey nest as one of their regular stops. Such universal appeal probably springs from many sources: the Osprey's potential tameness, its dramatic dives for fish, its monumental stick nests, its conspicuous breeding behavior, and its preference for the same sea and lake shores that people find so pleasing. Whatever the reasons, this fish-eating hawk has long delighted humans. And even though we are crowding their world with little thought of the consequences, Ospreys give every promise of being here to delight us in the decades ahead.

Despite this popularity, anyone hoping to learn about Ospreys has been forced, so far, to consult a multitude of different sources, most of them scholarly, narrowly focused, and out of date. This book aims to remedy that situation. In the chapters that follow, I have tried to synthesize an intimidating tangle of published and unpublished Osprey studies and to distill that material into a short, comprehensive, readable text, one free of jargon and accessible not just to the trained biologist but to anyone with an interest in birds, wildlife, or coastal ecology. The time seems especially ripe for such a synthesis. As a glance at this book's bibliography shows, Osprey studies have proliferated dramatically during the past 20 years, mostly in response to threats encountered by US and European breeding populations. Thanks to these studies, we now know more about Ospreys than we do about most other birds of prey, indeed most other wild animals.

But there is still a great deal to learn. This book will have succeeded if it stimulates new and innovative studies of Ospreys, especially of populations ignored by previous research. In addition, I hope this writing will introduce many to the pleasures of Osprey watching and to a new and better appreciation of the species.

In writing this book, I have depended heavily on outside sources, both published and unpublished. This is partly because the world's Osprey populations are so varied, so dispersed, and so numerous that no one person can ever hope to know them well. Despite the enjoyable days I have spent watching Ospreys in coastal Honduras, Kenya's Rift Valley, the Highlands of Scotland, and the boreal forests of northern Canada, my own field studies, and thus my real experience with this species, have been confined to just two regions of the eastern United States. One of these regions is Florida Bay, a rich mosaic of mangrove islets and shallow-water seagrass flats at the southern tip of the Florida Peninsula. The other region is an equally fertile ribbon of salt marsh bays and estuaries – the coastline of southern New England and eastern Long Island, NY. In addition, my own field research has focused mainly on Ospreys at the nest: their breeding behavior, reproductive ecology, and population dynamics. In order to write about Ospreys that I have never seen, therefore, and to illuminate facets of the Osprey's life history that I have never experienced first hand, I have borrowed extensively from other sources.

A few of these outside sources have been particularly helpful. Cramp & Simmons's (1980) scholarly treatment of Palearctic populations is a brief yet indispensable reference for anyone interested in the species, due mainly to contributions by Yves Prevost, an enthusiastic student of Ospreys during the 1970s. Osprey foraging ecology would be difficult to discuss without Prevost's (1977, 1982) two well-wrought theses, the more recent of which is the only comprehensive study of migratory Ospreys on their wintering grounds. Sten Österlöf's (1977) analysis of Osprey migration remains the definitive work on that subject, just as Paul Spitzer's (1980) research does in the field of Osprey population dynamics. Stan Wiemeyer has provided much of the data on contamination in Ospreys. Chuck Henny's prolific studies of Osprey nesting distribution in the United States and Mexico are a model of thoroughness and a tribute to what perseverance, good pilots, and a little government support can do. Finally, the proceedings of the Williamsburg, Montreal, and Sanibel Osprey conferences are key reference works (Ogden, 1977; Bird, 1983; Westall, 1984). The editors responsible for those proceedings, and others forthcoming, are to be congratulated for extracting so many good papers from Osprey researchers – a subspecies of biologist notoriously lazy about publishing.

Many people have helped with this book and the research that preceded it. I am particularly grateful to Paul Spitzer, who introduced me to Ospreys 14 years ago and who remains a ready source of ideas on these birds. My parents, Ralph and the late Kitty Poole, encouraged childhood learning with books, travel, and a farm to roam at will. Summers spent with Helen Hays's Great Gull Island Tern Project crystallized my interest in natural history and coastal environments. And five pleasant years as a graduate student in Woods Hole, Massachusetts, under the able guidance of Dr Ivan Valiela of the Boston University Marine Program, gave me the training and confidence to write this book. Roger Pasquier started me on this venture by introducing me to editors at Cambridge University Press, including my chief editor, Martin Walters, who patiently, tactfully saw this project through to completion.

Dozens of people have supplied data for this book. My gratitude extends beyond the brief notations made in the text. G. Clancy, R.D. Barradas, Roy Dennis, Tom Edwards, Mats Eriksson, M.D. Gallagher, R.E. Green, Erick Greene, John Hagan, Frederick Hammerstrom, Chuck Henny, P.A.D. Hollom, Bob Kennedy, Yossi Leshem, Tjelvar Ödsjo, P. and J. Olsen, Sten Österlöf, Jan Reese, Larry Rymon, Mike Scheibel, Sheldon Severinghaus, Charles Sibley, Paul Spitzer, Jon Swenson, S.P. Wetmore, and Guangmei Zheng deserve special thanks. Many of the photographs used herein were generously given. I am especially grateful for the photos supplied by J.R. Bider, Ms Eloise Beil (who led me to the excellent photos of the late H.H. Cleaves in the archives of the Staten Island Institute of Arts and Sciences), L. Cupper, Yossi Eshbol, Dr D.G.W. Hollands, Gordon Lind, Michael Male, Barry McCormick, B.-U. Meyburg, Sergey Postupalsky, E. Saïller, J.-F. & M. Terrasse, Joe Witt, and V. Serventy. Jack Cook's drafting skill is obvious in the diagrams included here, and Margaret LaFarge's graceful pen and ink drawings add immeasurably to the text.

A few people gave generously of their time in reviewing early drafts of this manuscript: Ken Parkes, Ivan Valiela, Betsy Bang, Tim Williams, Paul Colinvaux, Jon Swenson, Sandy Moss, Keith Bildstein, Ian Newton, I.C.T. Nisbet, and Stan Wiemeyer. In addition, Don Bourne, Judy Fenwick and Sylvia Sullivan kindly read late drafts of the entire manuscript and made countless helpful suggestions. All of these people improved the accuracy of the text and no doubt saved me from embarrassment. Remaining errors, however, are mine alone.

For help in the field, I would particularly like to thank Bev Agler, Gene Botelho, Topher Dudley, Vinnie Durso, Todd Highsmith, Tasha Kotliar, Ms E. Perkins, Nina Pierpont, and Mark West. Gil and Jo Fernandez kindly shared data with me and provided access to an established Osprey population. My Florida Bay studies were carried out under the auspices of the South Florida Research Center (Everglades National Park), with support from Audubon Research and Mr Herman Lucerne. Financial support for my research has come from the NY State Department of Environmental Conservation (Endangered Species Unit), the Mashomack Foundation, the National Science Foundation, and the Island Foundations.

I live near two superb research libraries, one at the Marine Biological Laboratory (MBL), Woods Hole, and the other at the Museum of Comparative Zoology, Harvard University. Both were indispensable while I researched this book. I owe special thanks to MBL's Jane Fessenden and Judy Ashmore for prompt and efficient library help. MBL also awarded me a three-month writing fellowship, the F.R. Brown Memorial Readership, and I am grateful for the quiet desk that came with that honor. Jim Broadus, director of the Woods Hole Oceanographic Institution's Marine Policy Center, likewise provided writing space for several months. Ethel Fenwick tended my daughter Phoebe daily during this past year, leaving me a quiet house in which to work. My gratitude to her, and to my wife Judy Fenwick, runs deep. Both tolerated my hermit writer's existence with better humor than I did.

Woods Hole, Massachusetts
Waquoit, Massachusetts
September 1987

1 Introduction

Osprey, the special one,
fisherman of the sea,
He does not have nets, he
does not beg for fish,
And he only eats fat fish,
The fisherman and his
boat,
The Osprey and his
skills,
There will be no lack of
fish.

(Wolof song, Senegal,
West Africa; from
Prevost, 1982)

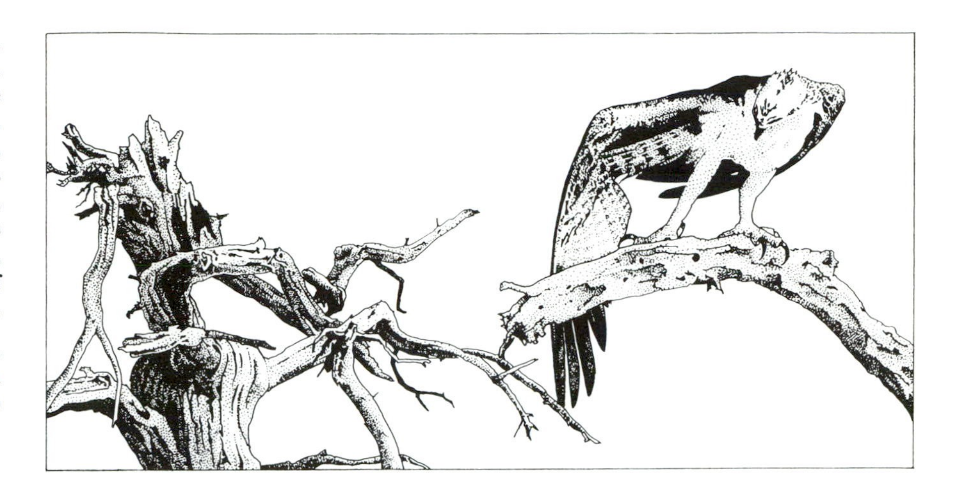

The first Ospreys I remember seeing were nesting atop tall, windblown trees that edged a narrow coastal pond, a serene sliver of wilderness in southern New England. It was a hazy summer afternoon as I waded the warm, murky margins of that pond, suitable weather in which to absorb my primeval surroundings. Harmless water snakes, startled from shoreline sunning spots, rippled past in bursts of energy. The unmistakable scent of beach roses drifted down from nearby dunes and a string of young teal scuttled ahead out of sight.

Ospreys were everywhere. Some perched beside huge stick nests that generations had used before them, and their slow whistled guard calls cast a dreamy spell over the surroundings. Others, generally males, lugged fish back from offshore waters, their prey locked tightly beneath them in their talons. As these birds approached their nests, small heads popped into view over the rims – nestlings anticipating a meal. Climbing to these nests as part of my routine surveys, I had an Osprey's perspective on the landscape, a wilderness coastline laid out below me. Here, I realized, were Ospreys as they always had been, here were nests like those that had overlooked Indian camps 10 000 years before.

A few days later, I checked another Osprey nest, not far away but in a very different setting. This nest topped a telephone pole along the edge of a small but busy commercial airport. Instead of wading, I was approaching this nest in a truck, equipment normally used by the local electric company to maintain their lines. Commuter planes droned overhead, car horns honked on a nearby highway, and floodlights, installed for night-time safety at the airport, stood ready to bathe the runway and nest in a sea of light. Yet the parent Ospreys perched nonchalantly nearby, only showing alarm when our truck stopped below them. Boosted aloft by the truck's hydraulic arm, I made a quick survey of the nest's contents. Two plump, healthy young, ready to fledge, crouched in the nest cup. It was a typically successful brood at this site.

What struck me then, the memory of wilderness nests still fresh in my mind, was how remarkably tame and adaptable these hawks were. Here was a large, wild bird of prey, a species that had evolved in a variety of wilderness habitats, living calmly alongside the crush of late twentieth century Americans. At a time when many of the world's raptors were increasingly threatened, here, it seemed, was a survivor.

As I came to know Ospreys better, this perception of resilience and adaptability was only strengthened. That airport nest was not an isolated oddity. Hundreds of Ospreys were nesting then (as now) in other developed sites: atop buoys and channel markers in busy commercial harbors, atop power poles (Fig. 1.1), radio towers, and lighting structures along major highways, and atop specially built nesting poles in dozens of suburban backyards.

Even without the benefit of such artificial nesting sites, the Osprey's broad, nearly worldwide distribution is impressive. Natural (tree or ground) nests (Fig. 1.2) occur in a great variety of habitats: along spruce- (*Picea* sp.) rimmed lakes of northern Canada, Siberia, and Scandinavia, on desert islands off Mexico, Oman, and Australia, along the heavily forested coasts of California and New Guinea, in cypress (*Taxodium* sp.) swamps of the coastal Carolinas, on mangrove islets in the Caribbean, along reservoirs in the western United States – the list goes on and on. Breeding Ospreys, in short, have managed to colonize much of the northern hemisphere's temperate and subtropical latitudes, as well as much of coastal Australia and nearby Pacific islands. Since most northern Ospreys winter in Latin America, Africa, and Southeast Asia, no continent except for Antarctica is foreign to this bird.

Ospreys have also survived many threats. In New England and the midwestern United States, populations bounced back recently after

Figure 1.1. Poles and pylons supporting utility lines provide ready-made nest sites for Ospreys, although birds occasionally are electrocuted (Chapter 10). This nest is in northwestern Mexico. (Photo: D.W. Anderson.)

the ban of pesticides that had nearly eliminated them. And in Great Britain, Ospreys have recolonized old haunts in recent decades after shooting and nest robbing had extirpated them from that country. Admittedly, other threatened Osprey populations have fared less well. But in an era when discouraging stories about the environment abound, when exploding human populations render our planet increasingly uninhabitable for many wild animals, the Osprey's tenacity provides real encouragement. Little wonder that this species has become so popular, and such a powerful totem for conservationists.

The Osprey is a large bird of prey, about the size of a small eagle or a large soaring (*Buteo*) hawk, although narrower winged than either of these. Seen in flight from below, the Osprey's salient features are its white or slightly mottled underparts, the pronounced crook in its long wings (they bend back at the 'wrists' or carpal joints), and its dark carpal patches (Figs. 1.3. and 4.1). Although Ospreys do soar

Figure 1.2. Osprey nest in a typical natural site. Dead trees, such as this one on eastern Long Island, NY (USA), are preferred nest sites. (Photo: A. Poole.)

high on fixed wings when conditions permit, most of their flight is active. Individuals row along steadily or hover over feeding spots like giant kestrels.

Seen perched from a distance, it is usually the Osprey's white breast that first catches the eye (Fig. 1.4). One quickly develops a search image for that gleam of white feathers. Seen up close, the Osprey's bright yellow eye, mottled breast band, and chocolate brown back and eye stripe are most prominent. Males sometimes lack breast mottling, are generally smaller than females and more buoyant on the wing. Both sexes, when perched, have an almost feline look about them, an impression enhanced, perhaps, by their small,

Figure 1.3. A female Osprey landing at her nest. (Photo: D. Haney.)

narrow heads. Recently fledged juveniles resemble adults except for their speckled backs and wings.

You hear Ospreys as often as you see them. Calls vary considerably (section 6.7), but a slow, whistled guard call (*kyew kyew kyew*) is heard frequently, especially when other Ospreys are nearby.

Handling an adult Osprey, the immediate impression is of sinewy strength and quickness, the fierce yellow eye alert to your every move. Ospreys seldom bite but their talons are sharp, quick, and unforgiving; one careless slip, and the handler may have to extract clenched talons from his arm. Female Ospreys, if their heads are hooded, are usually quiet and trusting in the hand. Males are generally much more highly strung, chittering and struggling even when hooded. Part of the excitement of handling these wild predators comes from knowing that each is a survivor, that each may have migrated thousands of kilometers every year of its breeding life. These are birds as familiar with tropical reefs and rain forest rivers as with northern lakes and estuaries. In many ways adult Ospreys are the equivalent of our professional athletes – superbly conditioned, seemingly inexhaustible, and the product of intense selection. Both travel widely, but Ospreys, of course, must provide their own transportation and navigation.

Unlike some birds of prey, Ospreys fare poorly in captivity. They resist life as trained hunters, shunning the falconer's wrist. I can find no reliable record of a captive Osprey retrieving fish for its human

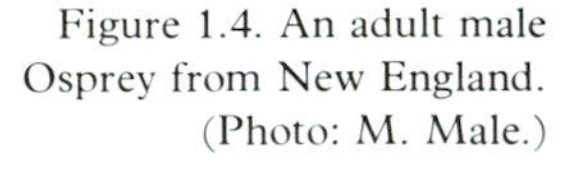

Figure 1.4. An adult male Osprey from New England. (Photo: M. Male.)

master; indeed, no other fish-eating bird of prey responds well to falconers either. Nineteenth-century Scottish farmers, however, circumvented this problem with typically frugal insight. They tethered wild fledgling Ospreys below nests and then snatched fish from the hungry young after the parent birds had made their deliveries.

Ecologists speak of an 'edge effect', the concentration of animal life that often occurs where two different habitats abut (Odum, 1971). Ospreys are part of this phenomenon, a shore and coastal bird, their life woven between land and water. Rarely do Ospreys nest more than a few kilometers from a river, lake, or sea. Even when not nesting, these hawks linger at the water's edge, perching openly or patrolling bays and shorelines for food.

Fish Hawks, as some of my older neighbors call Ospreys, are the only diurnal bird of prey that feeds exclusively on live fish. Like other hawks, Ospreys plunge feet first when hunting, but they cannot reach much below the water's surface. Thus fish are most available to them in shallow waters – bays, flats, river mouths, estuaries, along beaches and shorelines – just the sort of habitat where fish feed and spawn abundantly. It is no accident, therefore, that Ospreys congregate where shallow waters abound. The salt-marsh bays and estuaries of the eastern United States, the coastal lagoons of northern Latin America and Africa, and a few of the larger, shallower lakes of Scandinavia and western North America are particularly favored habitats. All are highly productive ecosystems where nutrients and the sun's energy are available throughout the water column. All, in other words, are particularly efficient at transforming sunlight into Ospreys.

Whatever the insights of ecology, Ospreys have come to mean more to us than just a terminal link in an ecosystem. Aldo Leopold (1949), North America's late, eloquent conservationist, spoke of the *numenon* of various habitats, that intangible essence different from *phenomena*, things which are ponderable, predictable, and accessible to science. The Ruffed Grouse (*Bonasa umbellus*), for instance, is a noisy, startling, conspicuous inhabitant of America's northern forests. Leopold saw it as the numenon of its habitat. Even though this grouse represents just a small fraction of the measurable energy flowing through its ecosystem, 'subtract the grouse and the whole thing is dead,' Leopold wrote. Ospreys provide the same sort of motive power to their habitats. Slow, whistled guard calls drifting over a foggy marsh; prominent stick nests dotting the tops of sentinel pines or rocky headlands; the burst of spray as a plunging Osprey hits the water – all lend energy to landscapes as we observe them, energy that cannot be measured in calories or filed in a computer.

For those more concerned with measuring and describing the Osprey's life, people who have made the writing of this book possible, the Osprey is an ideal study animal. Large size, open nesting and feeding habits, and potential tameness all render it delightfully easy to watch in most cases. In carrying out my own behavioral studies, for example, I wander through a neighbor's backyard to a riverbank, set up a folding chair and telescope, and settle down to note-taking. Ospreys nest just a few hundred meters away on low, marshy islands. No blinds or long distance treks are necessary; Ospreys are accustomed to seeing people where I sit. Not all Ospreys are this tame, but most will habituate to people given time (Chapter 9).

Where food and nest sites are abundant, Ospreys often breed in loose colonies. Such colonial habits facilitate behavioral observations and the collection of reproductive data. In my study area, I can easily check the contents of about 50 nests in one day, and naturalists in perhaps a dozen other regions can do almost as well. Many Ospreys nest less accessibly, but compared with most other raptors, birds which generally nest far apart and secretively, Ospreys provide enviably large sample sizes.

Another advantage of studying Ospreys is their diet. Fish come in neat, quantifiable packages, easily identified through a telescope or binoculars. Determining an Osprey's diet and food consumption is thus simpler than with most other birds. In addition, Ospreys hunt prey openly, so successful and unsuccessful hunts can usually be distinguished at a glance.

A final advantage of studying Ospreys is the ease with which individuals can be marked, generally with colored leg bands, and identified after release. This means that each bird's breeding and hunting success, its mate and nest site fidelity, its movements and longevity, and many other behaviors can be followed over time. Individual differences in behavior and competence provide a fascinating angle to any study of wild animals living under natural conditions. For it is on just such differences that selection operates and the evolution of a species proceeds. Although this book deals often, by necessity, with averages, I try also to emphasize differences among individuals and populations, differences that are obvious to anyone who has spent time with Ospreys. Like other complex animals, these birds are far from homogeneous. Even neighbors differ noticeably in behavior.

People, highly individual themselves, are just beginning to appreciate the individuality found within other species, a fact that Donald Griffin (1984) emphasizes in his provocative book, *Animal Thinking*:

> If one wishes to understand the behavior of animals, or still more if one is interested in their thoughts and feelings, one must take account of their individuality, annoying as this may be to those who prefer the tidiness of physics, chemistry, and mathematical formulations.

Before examining differences in Osprey habitat, behavior and reproductive success, we shall define the species more carefully, considering how these birds differ morphologically from other birds of prey and among themselves.

2 PHYLOGENY AND CLASSIFICATION

Every species is the terminus of an ancient lineage that has been hammered and shaped into its present form by a complex interplay of genetic recombination and natural selection. In a purely technical sense the resulting genome is richer in content than a Caravaggio painting, a Bach fugue, or any other great work of art.

E.O. Wilson (1985)

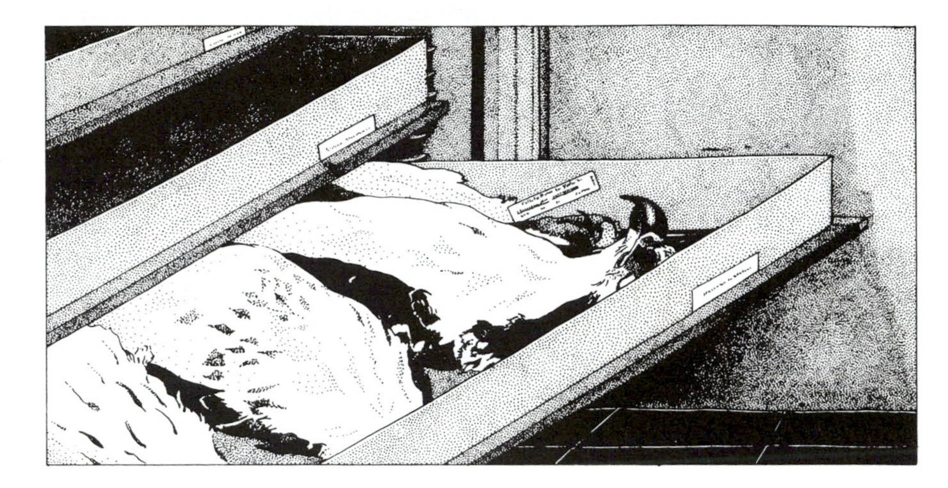

The Swede Linnaeus, who launched our present system of zoological nomenclature with his 'Systema Naturae' in 1758, named the Osprey *Falco haliaetus*, placing it in the same genus as most other birds of prey including the falcons and eagles. In 1809, Savigny created the modern genus *Pandion* for the Osprey, although he changed its specific name to *fluviatilis*, from the Latin '*fluvius*' (river), apparently because the specimens he examined were collected along the Nile. (In Savigny's time, there were no rules governing zoological nomenclature, and authors were free to follow previous writers or to coin names of their own. Later it became the custom to use the earliest species name that had been applied to an animal.) Most zoologists agreed that the Osprey deserved its own genus, so Savigny's generic name *Pandion* was universally accepted but coupled with the original Linnean species name *haliaetus*. This combination, still recognized today, was first used by Lesson in 1828.

Haliaetus is from the Greek '*hals*' (salt or the sea), presumably a reference to the Osprey as a bird of marine habitats. The second half of the specific name is from the Greek '*aetos*', eagle. Pandion was a legendary king of Athens whose relationship to Ospreys is tenuous. As E.A. Choate (1985) retells the myth, Pandion's two daughters, Procne and Philomel, had the misfortune to meet Tereus, king of Thrace. Tereus married Procne but soon decided that Philomel was more attractive. He hid Procne, told everyone she was dead, cut out her tongue to keep her quiet, and married the unsuspecting Philomel. Procne, dumb but undaunted, wove her story into a tapestry, which she sent on to Philomel. The sisters combined forces and vengefully roasted Tereus's only son, serving him up to his father for dinner. Even the gods drew the line at such behavior. For penance, they transformed Procne into a swallow, Philomel into a nightingale, and Tereus into a hawk, doomed to chase the other two eternally. Thus, as Choate points out, were Ospreys doubly misnamed:

> It seems that Savigny made two errors: First, if he wished to commemorate the mythological hawk, he should have named the new genus *Tereus*, as it was he who was metamorphosed, not Pandion. Second, it is rather stretching things ornithologically to have an Osprey that preys on fish chasing a swallow and a nightingale.

Since Linnaeus, most taxonomists have allied the Osprey with other birds of prey (order Falconiformes), but its closest relatives and proper taxonomic rank remained in doubt until recently. This uncertainty was due largely to changing taxonomic methods and disagreement over how best to determine relationships among species of birds. On the separate bases of osteology (skeletal structure), pterylosis (the distribution of feather tracts), and pelvic musculature, for example, various taxonomists of the early twentieth century all concluded that Ospreys were sufficiently different from other raptors to merit a separate family or suborder, but there was no consensus on their near relatives (Prevost, 1983a). Brown & Amadon (1968), in their comprehensive treatment of the world's hawks and eagles, placed *Pandion* in its own family within the suborder Accipitres (hawks, Old World vultures, and eagles) and linked the genus to the kites through shared morphological characters.

2.1 Biochemical evidence

Taxonomy based on morphology alone, however, can often be misleading. People seldom agree on which characters are the key ones. Because muscle, bone, and other body parts can be rapidly altered by the forces of natural selection, bodies that look similar to us may reflect convergence – the result of similar ecological pressures on unrelated species – rather than common ancestry. Taxonomists, therefore, have turned increasingly to biochemical techniques to unravel evolutionary history. Biochemical evidence provides a glimpse of the underlying genetic structure of an organism, which tends to evolve at a fixed rate and is less affected by the pressures of ecology (Barrowclough, 1983).

Comparing the chemical composition of the egg-white proteins of different raptors, Sibley & Ahlquist (1972) agreed with earlier conclusions that the Osprey deserved a separate family in the Falconiformes, closest to the Accipitridae, and that their morphological peculiarities could be explained as adaptive responses to their unique raptorial fish-eating niche. Recently, however, on the basis of extensive DNA–DNA hybridization studies, these two researchers amended their earlier position and reduced the Osprey to a subfamily (Pandioninae) within the family Accipitridae (Sibley & Ahlquist, unpublished). DNA–DNA hybridization is a method that offers heightened taxonomic precision by measuring the degree of association between single-stranded DNA molecules of different species, thereby comparing extensive segments of different genomes directly rather than comparing smaller segments of genomes indirectly by looking at their products – proteins and body parts (Sibley & Ahlquist, 1986). Despite significant morphological differences between Ospreys and other raptors, therefore, it appears this species is more closely related to the eagles, hawks, and Old World vultures than most earlier taxonomists had suspected. Which of these are the Osprey's closest relatives remains to be determined.

2.2 The fossil record

Fossils provide a few clues about the age of Ospreys and their early distribution. Harrison & Walker (1976) described an Osprey (*Paleocircus cuvieri*) from the late Eocene (about 50 million years ago) in Europe, but they based their description on somewhat tenuous evidence: a few claws similar in shape to those of modern Ospreys. Warter (1976) questioned this record and, based on more complete remains of wing bones, suggested that the oldest Osprey was his newly described *Pandion homalopteron* of the mid-Miocene of California, about 13 million years ago. Other mid to late Miocene fossils include an egg from Austria (Rauscher, 1984) and hind limb bones from Florida (Becker, 1985), so Ospreys were apparently well distributed in the northern hemisphere 10–15 million years ago. The

Florida fossils, remains of paleospecies *P. lovensis*, were complete enough to show that late Miocene Ospreys were quite similar to modern specimens, although not so robust. Other remains of Ospreys were found at about 12 Pleistocene sites (2–0.1 million years ago) in western Europe, North America, and the Bahamas (Brodkorb, 1964), a range essentially unchanged from that of the earlier Miocene.

Thus Ospreys quite similar to the ones we know today were well established in much of their current breeding range at about the same time our earliest ape-like ancestors left forests and began to walk upright across the plains of Africa. No doubt these primates saw Ospreys, migrants from the forests of Europe; perhaps they learned to recognize them. Certainly Pleistocene Ospreys were there to greet emerging humans when the latter pushed north into Europe. Björn Kurtén's (1980) fine novel *Dance of the Tiger* suggests that Ospreys played an integral part in the culture and myths of ice age Cro-Magnon and Neanderthal peoples. Ospreys would have shared boreal lakes and shorelines with both these subspecies of *Homo sapiens*.

2.3 Subspecies taxonomy

Sliding open a museum drawer full of dried Osprey skins can be a surprisingly enlightening experience. After recovering from the overpowering odors of fumigants and musky feather oil (the Osprey's preening gland secretes a pungent oil that lingers on the plumage for decades), one confronts rows of specimens from exotic corners of the world. Yellowed labels, some over a century old, dangle from dried legs, evoking an era when science and hunting were curiously mingled. Yet the first thing that strikes one is the similarity of the specimens; Ospreys look very much alike no matter where they come from. One of the more remarkable facts about this genus *Pandion* is how well it has resisted speciation, despite over 15 million years of evolution and a range that includes four continents. Subspecies have formed, to be sure, but taxonomists find none sufficiently distinct to merit status as a separate species.

To the layman, subspecies might seem an arbitrary designation, especially when calipers are needed to define differences. Yet subtle differences among races suggest a potential for species formation which time, isolation, and differing environmental pressures may eventually realize.

Four Osprey subspecies are currently recognized: *P. h. haliaetus* from the Palearctic (Europe, the northwest coast of Africa, and Asia north of the Himalayas); *P. h. carolinensis* from North America; *P. h. ridgwayi* from the Caribbean; and *P. h. cristatus* from Australasia (Australia, New Guinea, and nearby South Pacific islands) (Prevost, 1983a). Decades of arguing were needed to hammer out this classification. As new specimens were discovered and described during the nineteenth century, numerous taxonomists tried to make species of them, but these efforts generally failed. (The subspecies concept did not take hold until the end of the nineteenth century.) By the early 1920s, *Pandion* had been reduced to one species, composed of the four races recognized today. A few additional races were proposed, but none has proved worthy of recognition (Prevost, 1983a).

Size and plumage are what best separate the four Osprey subspecies, but the differences are not always straightforward. Finding geographic variation in Osprey body size, for example, is complicated by sexual dimorphism. Within subspecies, females not only weigh more than males but also have longer wings, tails, claws, and bills (Figs. 2.1 and 2.2). This means that male and female Ospreys from different parts of the world overlap considerably in size. Males from all populations nesting in the northern hemisphere are larger than Australasian females (Fig. 2.1). Yet in both sexes, Ospreys from tropical and subtropical climates tend to be smaller than those breeding at cooler, higher latitudes. Australasian Ospreys, for example, are 12–14% smaller (on average), sex for sex, than their Palearctic counterparts, and Red Sea Ospreys are slightly smaller than Swedish Ospreys (Fig. 2.1; Prevost, 1983a). Such shifts along a

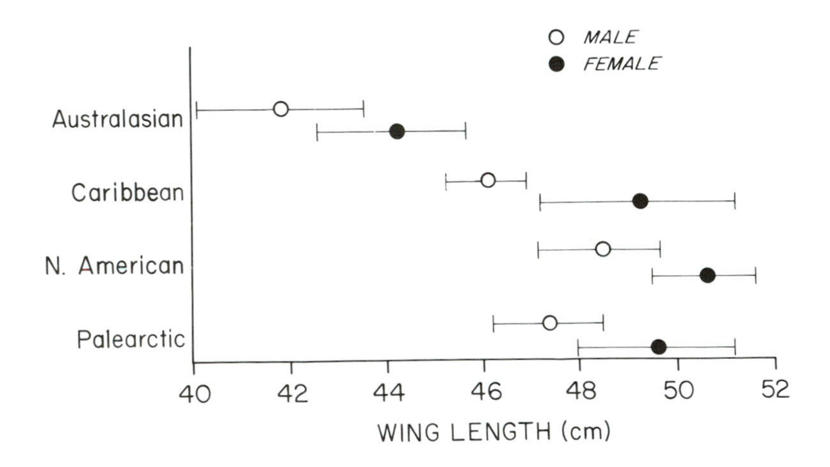

Figure 2.1. Mean wing length of male and female Ospreys belonging to the four subspecies recognized today. Means ± 1 standard deviation are shown. Wing length provides a good key to the overall size of these birds. Measurements, by Prevost (1983a), were from museum specimens preserved as skins.

climatic gradient are thought to have adaptive significance, following Bergmann's ecogeographic rule: among warm-blooded vertebrates, races in cold climates tend to have larger bodies than those in warm climates because larger bodies retain heat more efficiently.

Plumage provides some additional help in differentiating the world's Ospreys, but here again sexual differences confuse the issue. In Palearctic populations, females have slightly darker crowns than males, while in Australasia just the opposite is true (Fig. 2.3). And in all but Caribbean populations, females usually have fuller, darker breast bands than males do, although there is overlap between the sexes (Figs. 2.3 and 2.4). Relying on breast coloration alone, for example, I can separate males from females in only about 50–70% of the pairs I study in eastern North America. This percentage falls to less than 10% along Caribbean shores where the sexes have equally pale crowns and breasts; such paleness does, however, clearly separate this race from the other three (Fig. 2.3). No other race can be reliably distinguished by breast plumage alone, unless one compares birds of the same sex. One diagnostic character separating Osprey subspecies (irrespective of sex) is the pattern of feathers in the underwing coverts, but this is nearly impossible to observe in the field (Prevost, 1983a).

Despite sexual dimorphism, however, a combination of factors usually does let a field observer distinguish three of the four Osprey subspecies. *P. h. ridgwayi* is instantly recognizable by its very white head and breast plumage. Either sex of *P. h. cristatus* seems noticeably smaller than other Ospreys, and the combination of a dark

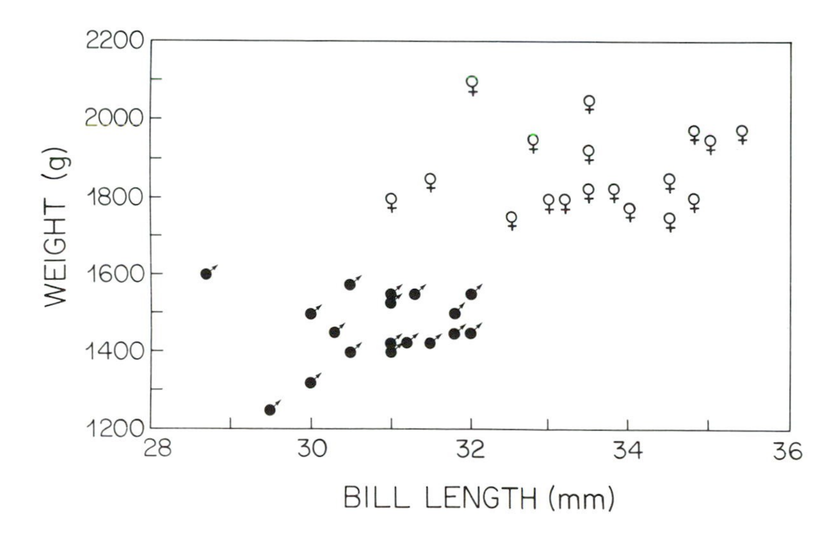

Figure 2.2. Bill lengths and body weights of individual male and female Ospreys (breeders) in New England. Because these two measurements are easy to take, they provide the best way to determine the sex of an adult Osprey in the field. (A. Poole, unpublished.)

breast band and a relatively light crown also set the Australasian birds apart (Fig. 2.3). *P. h. haliaetus* and *P. h. carolinensis* are the easiest to confuse, but the darker breast bands of the Palearctic birds (especially northern breeders) distinguish them in many cases. Southern Palearctic Ospreys, those breeding along the Red Sea and the Cape Verde islands, are paler than Ospreys from northern Europe and look much like North American individuals (Prevost, 1983a).

Why are Ospreys so similar? Why have they failed to speciate? Most other avian genera with as broad a distribution as *Pandion* are composed of numerous species. The *Haliaeetus* eagles (the Bald Eagle (*Haliaeetus leucocephalus*) and White-tailed Sea Eagle (*Haliaeetus albicilla*) being members of this genus), for example, form eight distinct species over a breeding range nearly identical to that of the Osprey (Love, 1983). Only the Peregrine Falcon (*Falco peregrinus*), among diurnal birds of prey, ranges as widely as the Osprey yet is also considered a single species (Cade, 1982). Northern Ospreys and Peregrines, however, are generally long distance migrants, sometimes dispersing hundreds of kilometers between natal and breeding sites. *Haliaeetus* eagles are more sedentary,

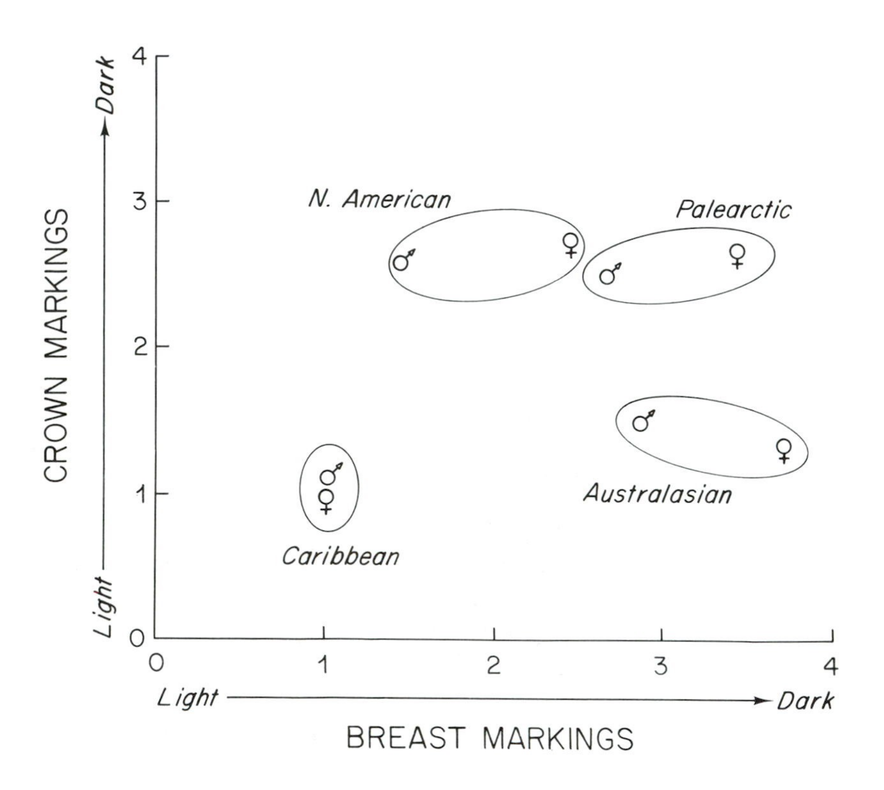

Figure 2.3. Plumage variation among the four Osprey subspecies; average values are shown. Breast markings were scored from 1 (almost white) to 4 (a wide band of dark feathers), crown markings from 1 (<10% dark feathers) to 3 (>50% dark feathers). Differences in plumage color alone are usually not enough to identify races reliably, especially when comparing birds of different sex. Data from Prevost (1983a).

mostly short distance migrants or year–round residents near breeding sites (Love, 1983). Thus local populations of these eagles are more likely to become isolated than populations of Ospreys and Peregrines. Carlos Wotzkow (1985), for example, noted several pairs of Cuban Ospreys in which one sex was the local *ridgwayi* and the other was *carolinensis*, and I have seen similar 'hybrid' pairs at nests in south Florida where *carolinensis* is the primary race.

Interchange among Osprey populations does occur, therefore, probably just often enough to help reduce isolation and the chances of speciation along with it. Prevost (1983a) speculated that climatic change during Pleistocene glaciations facilitated secondary contact between North American and eastern Palearctic Ospreys via the Bering Strait and that such gene flow might account for the similarities of *carolinensis* and *haliaetus*. Australasian Ospreys, on the other hand, are well isolated in the southern hemisphere (Fig. 3.5), and this may be why they have differentiated more extensively than others.

In summary, there are basically two Ospreys: an Osprey of

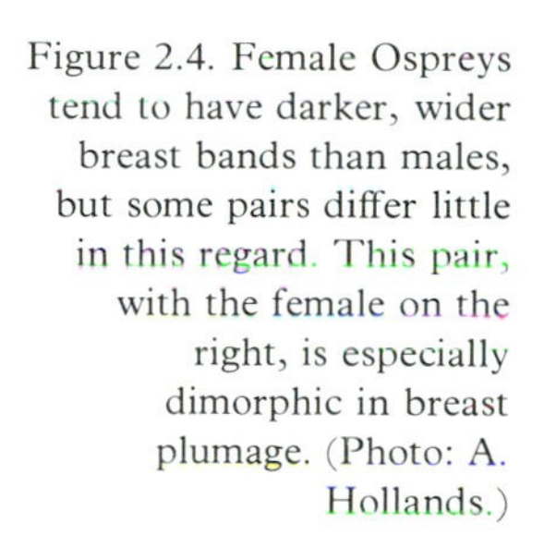

Figure 2.4. Female Ospreys tend to have darker, wider breast bands than males, but some pairs differ little in this regard. This pair, with the female on the right, is especially dimorphic in breast plumage. (Photo: A. Hollands.)

Holarctic regions (the north temperate zone), of which the Caribbean Osprey is a slightly modified form, and the Australasian Osprey. Size differences suggest the latter is the most ancient offshoot and that other resident populations are more recently formed races, but such differences could reflect isolation rather than antiquity.

3 STATUS AND DISTRIBUTION

The Osprey is a world citizen.

R.T. Peterson (1969)

In Africa . . . Ospreys are absent as breeding birds . . . All we can say is that it is inexplicable.

L. Brown (1970)

The study of Osprey distribution is a refresher course in world geography. Few birds are more cosmopolitan, more adaptable. Roger Tory Peterson (1969), in his relentless pursuit of the world's birds, has seen the Osprey's 'great nests of sticks in pine trees on Japanese islets, on sea cliffs near Gibraltar, on Sweden's spruce-rimmed Baltic coast, on pinnacles in Yellowstone National Park, USA, and on the headlands of Mexico's west coast.' A longer list of Osprey nesting territories provides a sampling of most of the coastal habitats of Europe, North America, northern Asia, and Australasia. For it is on these four continents that Ospreys breed, equally at home from subarctic to subtropical latitudes and beside river, lake, or sea. While many subtropical and Australasian Ospreys stay near nests year–round, north temperate Ospreys desert theirs each fall for warmer tropical wintering grounds. As either migrants or breeders, therefore, Ospreys have colonized all of the world's continents except for bleak Antarctica.

Despite this extraordinary range, three major gaps in the Osprey's breeding distribution are conspicuously evident: Africa, South

America, and Indomalaysia (India and southeast Asia). No evidence suggests that Ospreys ever bred in these regions, but they successfully winter there, so these gaps remain a puzzle. If we knew why such gaps occurred, we would know what governed the breeding distribution of this species. Answers to this puzzle are only touched upon in this chapter, which concentrates on documenting Osprey nesting abundance and distribution worldwide. Chapter 8 takes up in greater detail the questions of habitat suitability and barriers to population growth and dispersal.

Documenting the abundance and distribution of a species as widespread as the Osprey might seem a daunting task, but a surprising number of Osprey populations have been well surveyed. Even rough estimates of total numbers are rare for any species of bird, especially one that breeds on four continents, yet a few days' search through the published literature yields such an estimate for the Osprey: about 25 000 to 30 000 pairs, worldwide (Appendix 1 shows how this estimate was reached).

Osprey surveys have been effective for several reasons. First, these hawks are conspicuous nesters; bulky, treetop nests with large, white–headed birds perched nearby are hard to miss. Second, many Ospreys breed in loose colonies, so locating one nest can help the search for others. Third, Ospreys, especially those at temperate latitudes, tend to lay their eggs at about the same time as their neighbors do. Thus well timed surveys can often find most of a region's breeding pairs in just a few visits. Nevertheless, censusing the world's Ospreys has required real effort. The data this chapter summarizes were assembled by hundreds of people who cumulatively spent thousands of hours (often uncomfortable ones) driving back roads, wading swamps, and peering from small, cramped, bouncing airplanes.

The best surveyed Osprey populations have generally been small, remnant ones near settled areas where nests are visited each year; populations, often remote, censused by aerial surveys; and populations in affluent countries. But in almost every case, we know more about the geography and abundance of this species now than we did 25 years ago. Censusing efforts peaked during the 1970s, when declining numbers of Ospreys sparked concern for many populations.

Current numbers, however, are only half the key to a population's status. The other half is stability. To judge that accurately, one must also know a population's history, its past numbers. Unfortunately, most Osprey surveys before 1970 were anecdotal and subjective, making accurate definition of trends difficult. Only in Britain and the northeastern United States do we know the Osprey's history well (Chapter 11).

3.1 Palearctic Ospreys (*P. h. haliaetus*)

Migratory Ospreys nest in a broad, irregular pattern across the Palearctic, from Scotland east to the Kamchatka peninsula and Japan, and from treeline in subarctic regions south to about 40° N (Fig. 3.1). Most of these birds winter south of 20° N, in Africa, India, and southeast Asia, although some migrate only as far as subtropical latitudes (Fig. 3.1; Chapter 4). In addition to these migrants, a few remnant nonmigratory populations reside year–round south of 40° N along the coasts of the Mediterranean and Red seas, the Persian Gulf, the Cape Verde and Canary islands, southern Spain and Portugal, and possibly along lakes and rivers of central and western Asia (Fig. 3.1). Although European nesting populations have been increasingly well surveyed since the mid–1960s, the status and distribution of most Middle Eastern and Asian populations remain poorly known.

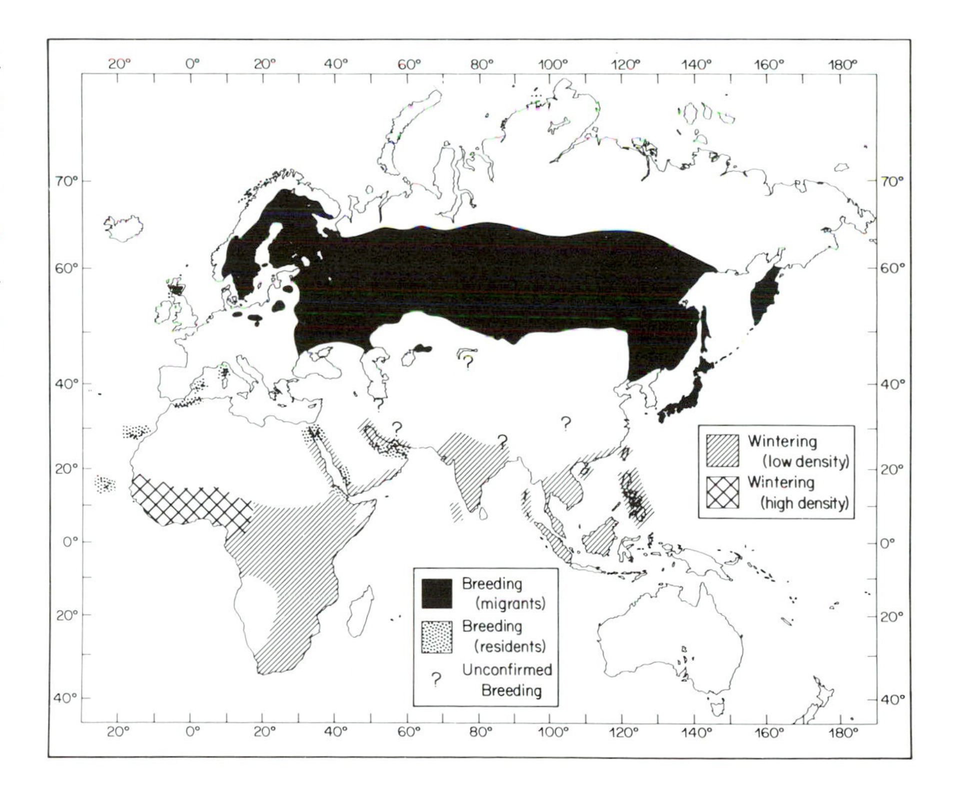

Figure 3.1. Breeding and wintering locations of Palearctic Ospreys, about 1985. Wintering range was determined mostly from band recoveries (section 4.1), breeding range from published sightings. Drawn from information in Cramp & Simmons (1980), Österlöf (1977), Prevost (1983a), and other sources mentioned in section 3.1.

All Palearctic populations have been persecuted to some degree, with their distribution influenced accordingly, but resident (nonmigratory) populations appear to have suffered the most.

Central and eastern Europe

Ospreys, it is thought, once nested abundantly in central and eastern Europe, but they are far less common there now than 100 years ago (Bijleveld, 1974). Both the shooting of breeders and the cutting of nest trees have taken a toll of these birds (Chapter 9). Nearly all active nests in this region are now confined to the lowlands of northern Germany and Poland (Meyburg & Meyburg, 1987; Fig. 3.1). In Poland, recent surveys revealed that 20 to 30 pairs of Ospreys nest annually, almost exclusively in the lake districts of Masuria and Pomerania (W. Krol, unpublished). Although Polish numbers are still declining slowly due to nest blowdowns and shooting, the increasing availability of artificial nest platforms, along with stronger penalties for killing raptors, should help to boost reproduction and stabilize this population.

The German Democratic Republic (East Germany) holds the largest share of central Europe's Ospreys, a total population of about 110 pairs (Meyburg & Meyburg, 1987). All of these birds breed in the pastoral lake districts of the north German lowlands, about 50 pairs in Brandenburg and the remainder to the north in Mecklenburg. Although portions of the Mecklenburg population disappeared during the 1960s, the loss has been balanced by a recent gain in numbers at Lake Muritz; there, 15–20 pairs nest together on a string of pylons supporting electrical lines (Fig. 3.2). Similar pylon nests are seen in Brandenburg. Over 50% of the Ospreys in these two regions now nest on pylons or at other artificial sites (Meyburg & Meyburg, 1987). Few if any breeders remain along the Baltic coast, so German Ospreys now depend almost exclusively on fresh water lakes while nesting. Given the high reproductive rates of these birds, however, the abundance of artificial platforms available to them, and the growing protection they enjoy, their future looks promising.

During the eighteenth, nineteenth, and early twentieth centuries, the Osprey was an uncommon but regular breeder in many other countries of central and eastern Europe: Austria, Switzerland, Czechoslovakia, West Germany, Belgium, France, and perhaps Hungary (Glutz, Bauer, & Bezzel, 1971; Bijleveld, 1974; Becsy & Keve, 1977; Cramp & Simmons, 1980). Today, these populations have nearly disappeared, probably due to shooting and to their small

Figure 3.2. Ospreys nesting on power poles in Mecklenburg, northern Germany. *Top:* arrows show individual nests. (Photos: B.-U. Meyburg.)

initial size (Glutz, *et al.*, 1971; Bijleveld, 1974). Historically, the last pair of Ospreys in mainland France bred in Lorraine in 1945; in West Germany, the last pair bred during the early 1960s in Westphalia and Bavaria. Recently, however, Ospreys have recolonized central France. One or two pairs now breed near ponds in Sologne, a lovely region of marshes, woodlots, and old hunting estates about 70 kilometers south of Orleans. This promising development suggests that Ospreys may soon recolonize other European countries, although reintroductions (section 10.3) would greatly speed this process.

Osprey numbers have also dwindled in the Balkan countries, although populations were always small here, at least during the last 150 years (Bijleveld, 1974; Cramp & Simmons, 1980). In Rumania, one or two nesting pairs are thought to remain in the Danube Delta and the Letea Forest (Kalaber, 1985). In Greece, Ospreys were last known to breed in 1966 in the Evros delta (Glutz *et al.*, 1971). In Turkey, a few pairs may still breed in Thrace and at other localities along the Black Sea (Beaman & Porter, 1985). None of these recent records has been confirmed, however, and all of these populations could well be extinct. Raptors are poorly protected in these countries.

In summary, the Ospreys of central and eastern Europe have declined significantly during the past 150 years, due largely to persecution. Their numbers are now stable at 130 to 140 breeding pairs, most confined to the lake districts of northern Germany and Poland. Nearly half these birds nest at artificial sites, which are increasingly available in this region. Such nest sites, and declining rates of shooting, should boost numbers and stimulate the recolonization of former ranges.

Fennoscandia

Despite sporadic persecution of Ospreys in Finland and Scandinavia, these regions remain strongholds for the species. Lack of human settlement and large expanses of undisturbed forest bordering shallow, productive lakes have undoubtedly helped Ospreys to thrive here ever since the glaciers retreated. Similar boreal forest habitat supports Ospreys throughout the Holarctic. Fennoscandian Ospreys are distributed relatively evenly compared with farther south where pairs often nest in colonies.

About two-thirds of Fennoscandia's Ospreys nest in Sweden where roughly 2000 breeding pairs were counted by members of the Swedish Ornithological Association during an ambitious country-

Figure 3.3. *Top:* Lake Åsnen, typical Osprey nesting habitat in southern Sweden. Ospreys favor such habitat because the water is shallow, making fish accessible, and because small islands are plentiful, providing ideal nesting sites safe from mainland predators. *Bottom:* One such islet, with a nest atop its lone pine. (Photos: J. Sondell.)

wide survey (much of it on foot) in 1971 (Österlöf, 1973). Most of these Ospreys nest in the southern third of Sweden, with the highest densities (1 pair / 5 km^2) on islands in Lake Marlen, south and west of Stockholm, and Lake Asnen (Fig. 3.3). North of 60° N, pairs are more scattered; north of 66° N, they are decidedly rare (1 pair / 200–300 km^2), indicating a pronounced north–south gradient in breeding density. Only 10–20% of the population breeds along the Baltic and west coasts, in estuarine environments – a curious pattern given the abundance of potential island nesting sites and the Osprey's concentration along similar coastlines of North America.

Finnish Ospreys are distributed much like those in Sweden, although they never reach such high densities. Pertii Saurola (1986) and co–workers started annual countrywide surveys in Finland in 1971 and these continue. Here again, most of the 900–1000 nests were found on foot, a tribute to the energy and perseverance of the census takers. They found maximum breeding densities of about five to seven pairs per 100 km^2, equivalent to densities at similar latitudes in central Sweden. This supports the suggestion that latitude determines Osprey breeding density in these subarctic environments, perhaps through food supply. A very few pairs pioneer north of the arctic circle into the spruce bogs of Finnish Lapland, the northern limit of the Osprey's range in the Holarctic. Forty percent of the Finnish population now nests on artificial platforms, mostly topped trees in areas where the birds have traditionally nested (Saurola, 1978 & 1986). Despite recent concern about the harmful effects of pesticides and acid rain, Swedish and Finnish Ospreys continue to show strong reproductive success and stable numbers (Österlöf, 1973; Odsjö & Sondell, 1976; Saurola, 1986).

Ospreys in Norway and Denmark have fared worse than their neighbors. Although we lack accurate historical numbers, Ospreys were common in Norway in the mid–nineteenth century. By 1940, persecution had reduced this population to only three or four pairs (Bijleveld, 1974). With increased protection since World War II, however, Norwegian Ospreys have recovered rapidly and now (1985) number between 150–200 pairs (A. Haga, unpublished). Some of these birds breed along coastal fjords but most are found at inland lakes, often nesting on islands (Haga, 1981).

In Denmark, recovery has been far slower (Bijleveld, 1974; Dyck, Eskildsen, & Moller, 1977). Ospreys bred there only sporadically from about 1900 to 1960, but in recent years a few pairs have returned to nest regularly at lakes north of Copenhagen. Efforts are being made to keep the exact nesting locations secret. Despite official protection, therefore, these birds are apparently still not safe. Given people's tolerance for Ospreys in the rest of Fennoscandia, it is surprising that Danish attitudes have progressed so slowly.

Mediterranean

Greek and Phoenician sailors plying the coastal waters of the Mediterranean probably saw Ospreys regularly. Today, this population is scattered and threatened, confined to just a few islands and remote stretches of coast: Corsica, the Balearic islands, and portions of Morocco and western Algeria. As this population has dwindled, its distribution and abundance have become increasingly well known.

Over 100 Ospreys are thought to have nested in Corsica during the early 1900s, but shooting and egg collecting reduced this number to three or four active nests by the late 1960s (Terrasse & Terrasse, 1977; Saïller & Nardi, 1984). Protection by the Parc Naturel Regional de Corse in the early 1970s has led to a modest recovery, about 10 pairs by 1980, which seems to be continuing. Most pairs nest at traditional sites on cliffs and pinnacles along the park's rocky coastline (Fig. 3.4; Bouvet & Thibault, 1981).

On the Balearic Islands, Ospreys are less well protected and faring poorly. Mounting development pressures, especially on tourist–packed Majorca, have cut this coastal population, probably small to begin with, to only six or eight pairs (Terrasse & Terrasse, 1977; Muntaner, 1981). These Ospreys may receive some protection from a proposed park but, overall, they appear seriously threatened.

Figure 3.4. Ospreys in Corsica and elsewhere along the Mediterranean nest on high, rocky promontories like this one. Corsica's red sandstone seacliffs provide particularly dramatic nest sites. (Photo: E. Saïller.)

Building artificial nest sites in undisturbed locations is perhaps the first management step needed to protect these birds.

The Berthons (1984) journeyed by rubber boat along the remote, rugged coast of Morocco and discovered 18 pairs of Ospreys west of Cap des Trois Fourches, a significant increase over earlier estimates for this region. Most nests were on cliffs or rocky pinnacles, with pockets of high density. One pair of Ospreys continues to breed on the Chaffarinas Islands (Witt *et al.*, 1983), near the Moroccan–Algerian border, and some ten to fifteen pairs have been estimated for coastal Algeria, most west of Algiers (Jacob, Jacob, & Courbet, 1980). In spite of proposed protection for many of the Moroccan nest sites and the inaccessibility of much of this North African coastline, the status of these birds is precarious and deserves continued monitoring.

Ospreys have disappeared from most other regions of the Mediterranean. Small numbers nested before 1900 in Sardinia and before 1955 in Sicily, but shooting and nest destruction eliminated the last of these (Bijleveld, 1974). Ospreys also bred along the Spanish coastline near Cadiz and Gibraltar and along the coast of southwestern Portugal, but these populations are now much reduced. Two nests remain active along Portugal's coast, four along the southeastern coast of Spain (L. Palma & L. Gonzalez, unpublished). Until hunting is better controlled in these Mediterranean countries, however, recovery will be slow and difficult. Ospreys are still shot there.

The Atlantic Islands

P. h. haliaetus breeds south into subtropical and tropical latitudes on the Cape Verde and Canary islands, off the northwest coast of Africa (Fig. 3.1). Early naturalists visiting these islands found Ospreys a conspicuous part of the local fauna – so conspicuous, in fact, that the birds were eagerly collected as scientific specimens (Bannerman, 1963). Nest robbing by local residents further reduced the population. In the Canaries, 13 pairs bred in 1985 (L. Gonzalez, unpublished), roughly the same number found there one to two decades earlier (Bannerman & Bannerman, 1968; Cramp & Simmons, 1980). In the Cape Verdes, Ospreys are apparently more numerous. About 50 pairs nested there during the 1960s, scattered along the cliffs and rocky shores of that impoverished archipelago (de Naurois, 1969). If these nests remain, the Cape Verdes hold more

Ospreys than the entire western Mediterranean, or Scotland. Clearly this population deserves further study and protection.

The Middle East

Ospreys are uncommon, year-round residents in the Middle East, breeding primarily on islands in the Red Sea and the Persian Gulf. Their range overlaps slightly that of wintering migrants from Europe (Fig. 3.1). Although Ospreys might seem an anomaly along this barren desert coast, one has only to slip into these waters with mask and snorkel to realize that fish are very plentiful here, especially along the coral reefs which fringe this coast. Where food and safe nest sites are readily available, climate is a minor barrier for the Osprey.

A stronghold of this resident population is Tiran Island, in the southern Gulf of Aqaba, where a colony of about 30 pairs nested during the 1970s (S. Suaretz & Y. Leshem, unpublished). Tiran lacks mammalian predators so Ospreys nest safely here on the ground, sharing this desert sanctuary with other rare nesting birds like Spoonbills (*Platalea leucorodia*), Sooty Falcons (*Falco concolor*), and various herons. Tiran was owned and protected by Israel up to 1982, when it passed to Egypt through peace accords. It is now under the authority of Egyptian and UN military forces and visitors cannot land, making the island a *de facto* nature reserve, off limits even to scientists (Zimmerman, 1984). Ospreys also breed on nearby Snapir Island and along the southern coast of the Sinai, bringing the total for the northern Red Sea to about 50 pairs (Y. Leshem, unpublished).

Elsewhere, Red Sea Ospreys are poorly known. The Dahalac Archipelago (off Eritrea – northern Ethiopia) harbored 20 to 30 pairs during the early 1960s (Clapham, 1964), but no one has surveyed this population since then. Likewise, little is known about Ospreys along the south coast of Egypt, the Sudan, Yemen, or western Saudi Arabia, all areas where the species may have bred locally during the first half of the twentieth century (Mackworth-Praed & Grant, 1957; Smith, 1957; Brown, 1970; P. Hollom & M. Gallagher, pers. comm.). A few Ospreys continue to nest along the north-east coast of Oman, especially around Musandan and on islands in the Strait of Hormuz (Gallagher & Woodcock, 1980). In other areas of the Persian Gulf, scattered pairs nest on islands off Bahrain, Qatar, and the United Arab Emirates, but no one has ever censused these birds (Bundy & Warr, 1980; Gallagher & Rogers, 1978). In northern Iran, Österlöf (1965) surveyed Ospreys nesting along the Caspian Sea and found this population reduced to just a few pairs. By now, the entire Iranian Osprey population has probably disappeared.

Clearly, the Middle Eastern Osprey population is small and vulnerable. Nest robbing and shooting continue. Accurate surveys are still needed to determine the status of much of this coastal desert population.

USSR

A glance at Figure 3.1 suggests that Ospreys are broadly distributed in the USSR, absent only from the most northern, southern, and western regions. Yet mapping Osprey distribution here is a difficult task; reports are few and the territory is vast. While the boundaries shown in Figure 3.1 are probably fairly accurate, many regions within those boundaries do not support nesting pairs.

Most Ospreys breeding in the USSR are confined to boreal lakes and rivers, habitat similar to that supporting the species in much of Canada and Fennoscandia. Galushin (in Cramp & Simmons, 1980) suggested that Ospreys were still common (two pairs per 100 km^2) in much of the northern USSR, but rare in western regions due to human pressures. Estonia, Latvia, and areas around Leningrad, for example, are now thought to hold only about 50 pairs each. Overall, the Osprey is listed as rare in the USSR. Official protection became stringent there only during the 1960s and 1970s (Galushin, 1977); enforcement may still be lacking.

Several recent reports from the 1983 Moscow Conference on Birds of Prey (Galushin & Flint, 1983) provide significant new data on the numbers, status, and distribution of Ospreys in the USSR, as follows:

(i) The Volga Delta (47° N, 47° E). A minimum of 20 nests were estimated for this region, most on small river islands of low, dense willow trees. 'Full protection' is given to Ospreys here, including their nest trees.

(ii) The Altai-Sayansky mountain region, 400–500 km east of Lake Baikal. People here report a few nests scattered throughout local nature reserves, including several nests on Teletsky Lake (52° N, 87° E) and several in the Yenesey Valley (52° N, 95° E).

(iii) The Par and Nadim rivers (67° N, 78° E) – a region of Siberian taiga forest, north of the arctic circle. Two individuals were sighted here in June and one nest was reported destroyed.

(iv) Northern Byelorussia (Vitebskaya Province, 36° N, 30° E). This region of lakes, rivers, and forested sphagnum bogs supports

> 50–60 nesting pairs, probably one of the densest breeding concentrations of Ospreys in the USSR. Some nests are only one to two kilometers apart. The question remains whether the USSR holds many other such pockets of density.

There are recent reports from two other regions of the USSR. Lobov (1985) estimated that 60 to 80 pairs of Ospreys breed on the Kamchatka peninsula (55° N, 157° E), a relatively undisturbed land where many other birds of prey find refuge. Ospreys have abandoned some areas of Kamchatka, but overall this population is thought to be stable. In the western Ukraine (48° N, 23° E), by contrast, Ospreys disappeared as breeders 50 to 100 years ago when forests were cleared (Gorban, 1985). It is mainly in less developed areas, therefore, that Ospreys have managed to survive in the USSR.

Japan and Korea

Although Ospreys are regular passage migrants through Korea, there is no record of the species ever having bred there. It is not clear if this absence is due to historical persecution or to a failure to colonize this region, but Korea does form a notable gap between Ospreys breeding in Manchuria and those breeding to the south and east in Japan.

In Japan, reports of Osprey are sketchy, but breeders seem to be very sparsely distributed. Isolated nests occur locally from Hokkaido south to Kyushu, most on cliffs or nestled in the tops of windswept pines along rocky coastlines (Anon, 1974). In particular, some of the smaller nearshore islands support breeders, perhaps because of reduced human pressures. Persecution can be a problem. The Wild Bird Society of Japan, which has surveyed some parts of the country for Ospreys, does not reveal nest locations for fear the birds will be harmed. Japan, after all, is a nation of fishermen and fish farmers, to whom Ospreys may seem a threat. Unfortunately, despite a few local surveys, no one has tried to census all of Japan's Ospreys. Japan has a strong interest in wildlife, especially large, wild birds. One might expect that this remnant Osprey population would be better known and protected.

India, China, Southern (Continental) Asia

Rumors persist that Ospreys nest along the lower reaches of the Himalayas in Ladakh, Kashmir, Garhwal, Kumaon, or Assam, but these rumors have never been substantiated (Ali & Ripley, 1968). Confusion with wintering birds or migrants is the likely basis for such reports. Similarly, reports have suggested that Ospreys once bred in southeastern China and on Taiwan (Vaurie, 1965; Cheng, 1976), but naturalists presently familiar with these regions have never found nesting pairs, even though Ospreys migrate and winter there (Z. Guangmei & S. Severinghaus, personal communication). Elsewhere in China and southeastern (mainland) Asia, breeding Ospreys are also absent (Fig. 3.1). It is curious that nesting pairs have never penetrated as far south in Asia as they have on the fringes of Africa and in the Middle East.

3.2 Australasian Ospreys: *P. h. cristatus*

Ospreys inhabit much of coastal Australia plus a broad arc of islands stretching north and east of that continent from New Caledonia to the Celebes and Java (Fig. 3.5). This is a nonmigratory population, well surveyed (for distribution, not numbers) only in Australia. Outside of Australia, much of our information on *P. h. cristatus* comes not from field observations but from a small number of museum specimens, most collected decades ago with no remarks on breeding ecology. Ornithologists with a sense of adventure, and perhaps a taste for Joseph Conrad's novels, should be encouraged to resurvey the Ospreys of these southwest Pacific Islands.

Australia

Australian Ospreys, exclusively coastal nesters, have colonized a wide range of latitudes from temperate South Australia to tropical Queensland (Fig. 3.5). During the nonbreeding season, a few of these birds wander inland along larger rivers but most remain at the coast (Blakers, Davies, & Reilly, 1984). Inland waters here may be too turbid or barren to support predictable fishing. Ospreys breed on coral islands along the Great Barrier Reef and on rocky islands and cliffs near Spencer Gulf and Adelaide in South Australia (Serventy, 1965; Cupper & Cupper, 1981). Western populations, separated from others by coastal deserts, form a few small breeding colonies on offshore islands (Fuller & Burbidge, 1981), but elsewhere nests are isolated in typical sites: dead trees, electrical pylons, rocky cliffs, and mangrove (*Rhizophora* sp.) swamps. The Australian Osprey population is currently stable, well protected, and largely free from contamination (P. & J. Olsen, personal communication).

Ospreys have always been absent as breeders in Tasmania,

southeastern Australia, and New Zealand, although it is not clear why this is so.

Southwest Pacific

Judging from sight records and museum specimens, Ospreys that nest on southwest Pacific islands are most numerous along the coasts and lowland rivers of New Guinea (Rand & Gilliard, 1968; Prevost, 1983a). North and west of this, *P. h. cristatus* is thought to breed on the Molucca and Lesser Sunda Islands, as far west as the Celebes and perhaps Java (Prevost, 1983a). Museum workers have suggested that Ospreys also breed in the Philippines (Delacour & Mayr, 1946; duPont, 1971), but the specimens examined were not identified as *cristatus* and no proof of nesting was given. Ornithologists with field experience in the Philippines have never found Ospreys nesting there (R. Kennedy, personal communication). The few individuals seen and collected in those islands were probably Palearctic migrants, just like the Ospreys found at similar latitudes in Sumatra, Malaysia, Borneo, and southeast Asia (Smythies, 1960; Medway & Wells, 1976).

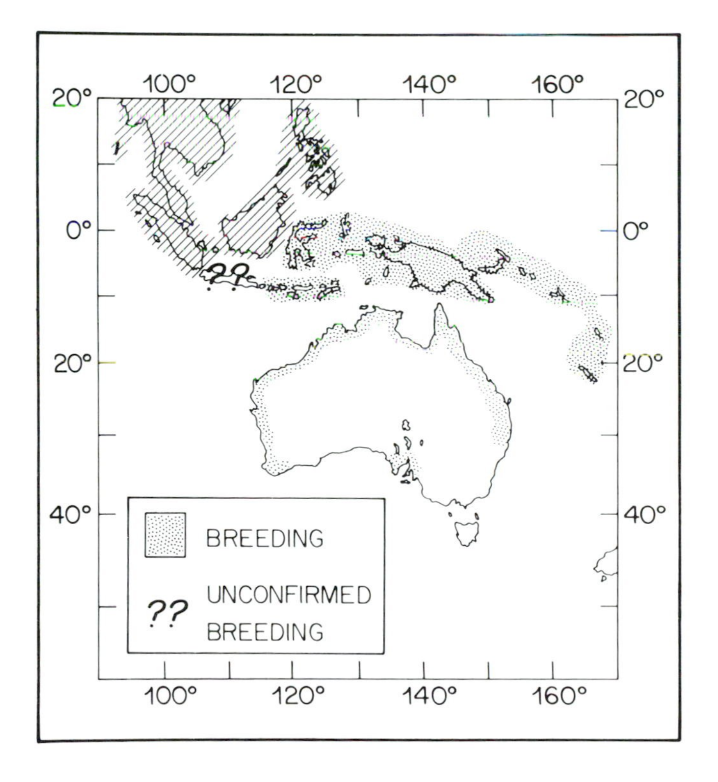

Figure 3.5. The known breeding range of Australasian Ospreys (*P. h. cristatus*), about 1980. These birds do not migrate. Diagonal lines show the wintering range of migrant Ospreys from the eastern Palearctic (*P. h. haliaetus*). Note the lack of overlap between the two subspecies. From Prevost (1983a) and Blakers *et al.* (1984).

South and east of New Guinea, Ospreys have colonized an arc of islands extending from the Bismark Archipelago to New Caledonia, including the Solomon Islands and Vanuatu (New Hebrides; Fig. 3.5). Specimens and sightings from this region are rare, however, and no reports exist to tell us how these populations are faring. Ospreys have not spread east of this region to Fiji or beyond, a curious gap given the ease with which these hawks cross water.

3.3 Caribbean Ospreys: *P. h. ridgwayi*

People have a good idea of where Ospreys nest in the Caribbean, even if they do not know how many remain there. Cuba, the southern Bahamas, and the east coast of the Yucatan Peninsula are the only known centers of breeding for *ridgwayi*, a nonmigratory race (Fig. 3.6). Recent reports from Cuba suggest that nesting Ospreys are scattered along much of that island's coastline, as well as on surrounding islets; mangrove trees and buoys are favored nest sites (Wotzkow, 1985). Wotzkow found Ospreys most numerous in eastern Cuba, but he observed only 28 nests on the entire island, many without breeders. His surveys were only sporadic, however; an effort should be made to census this key population during the height of the breeding season.

Using a small, low-flying airplane in 1971, Sandy Sprunt (1977) surveyed the east coasts of Mexico and Belize for wading birds and

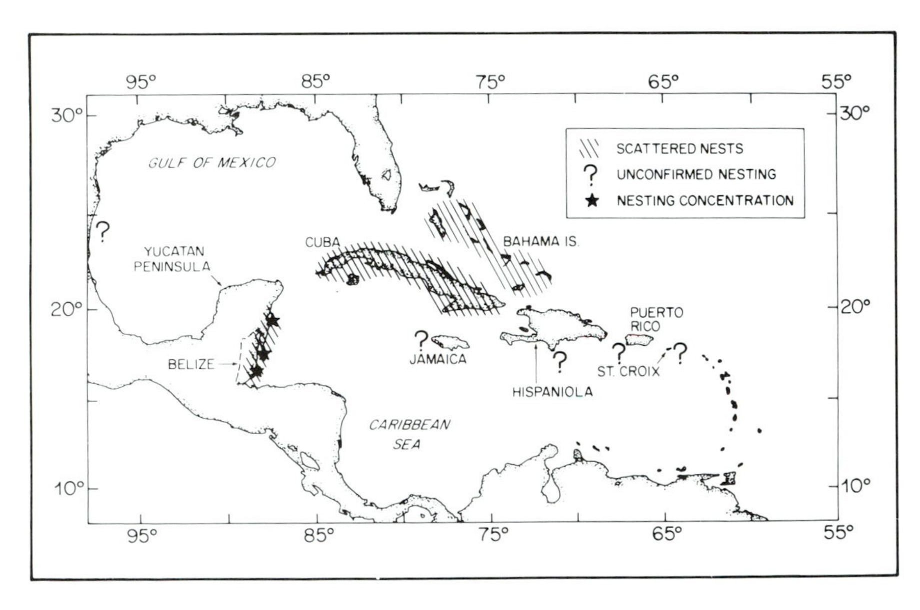

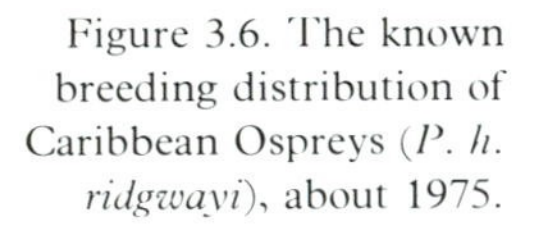
Figure 3.6. The known breeding distribution of Caribbean Ospreys (*P. h. ridgwayi*), about 1975.

Ospreys. He flew in early spring, saw many lone Ospreys (probably migrants) along the Gulf coast of Mexico, but located few nests until he reached the province of Quintana Roo, on the Yucatan peninsula (Fig. 3.6). There Sprunt counted more than 20 nests scattered along the shores of mangrove islets and fertile, shallow bays. Farther south, in Belize, he spotted another 20–25 nests on small coastal islands, including a significant colony of over 15 nests on the Turneffe Islands. The total breeding population of coastal Yucatan and Belize in 1971 may have been as high as 50 pairs. Holding perhaps 30%–50% of the world's remaining *ridgwayi*, this region urgently needs to be resurveyed. Much of this coastline remains undisturbed wilderness, but some areas have been developed recently for tourism. There is no record of how Ospreys are adjusting.

In other parts of the Caribbean, Ospreys are uncommon breeders. They are found scattered throughout the Bahamas but have disappeared from the Greater and Lesser Antilles, islands where they were probably rare to begin with (Wiley, 1984; Fig. 3.6). Rough estimates thus suggest that the entire Caribbean population may now number only 100–150 pairs (Appendix 1).

3.4 North American Ospreys: *P. h. carolinensis*

Ospreys depend on a wide range of North American habitats for breeding: rivers in Labrador, chains of boreal lakes in Alaska, salt marsh bays in New England, and the desert coastline of northern Mexico, to name just a few (Fig. 3.7). Although many of these habitats have similar ecological counterparts elsewhere in the world, North America is unique in the success with which Ospreys have colonized its shallow-water ecosystems. Given what we know from historical records, the range and density of this population has never been matched by other Ospreys.

Although historical data are sparse, recent surveys of North America's Ospreys have been remarkably thorough. Perhaps 60%–80% of the roughly 8000 active Osprey nests in the United States have been located at least once during the past 20 years. Surveys were especially prevalent during the 1970s when the threat of pesticides generated concern and funding. A combination of air and ground surveys, led by Chuck Henny of the US Fish and Wildlife Service, censused key breeding concentrations in the United States and Mexico. These surveys differed in accuracy, depending on the extent of ground coverage (Henny & Noltemeier, 1975), but together provided the first rough estimates of this huge population (Henny, 1983). Just as importantly, they helped to initiate a number of annual state surveys that have continued into the 1980s. In Canada and Alaska, where the territory is vast and nests are scattered, Ospreys have been less thoroughly censused.

Most North American Ospreys are migratory, wintering in Latin America and the Caribbean Basin with concentrations in northern South America (Fig. 3.7; Chapter 4). South of about 31°–33° N, in Florida and Mexico, the climate permits year-round occupancy and Ospreys nesting there do not migrate.

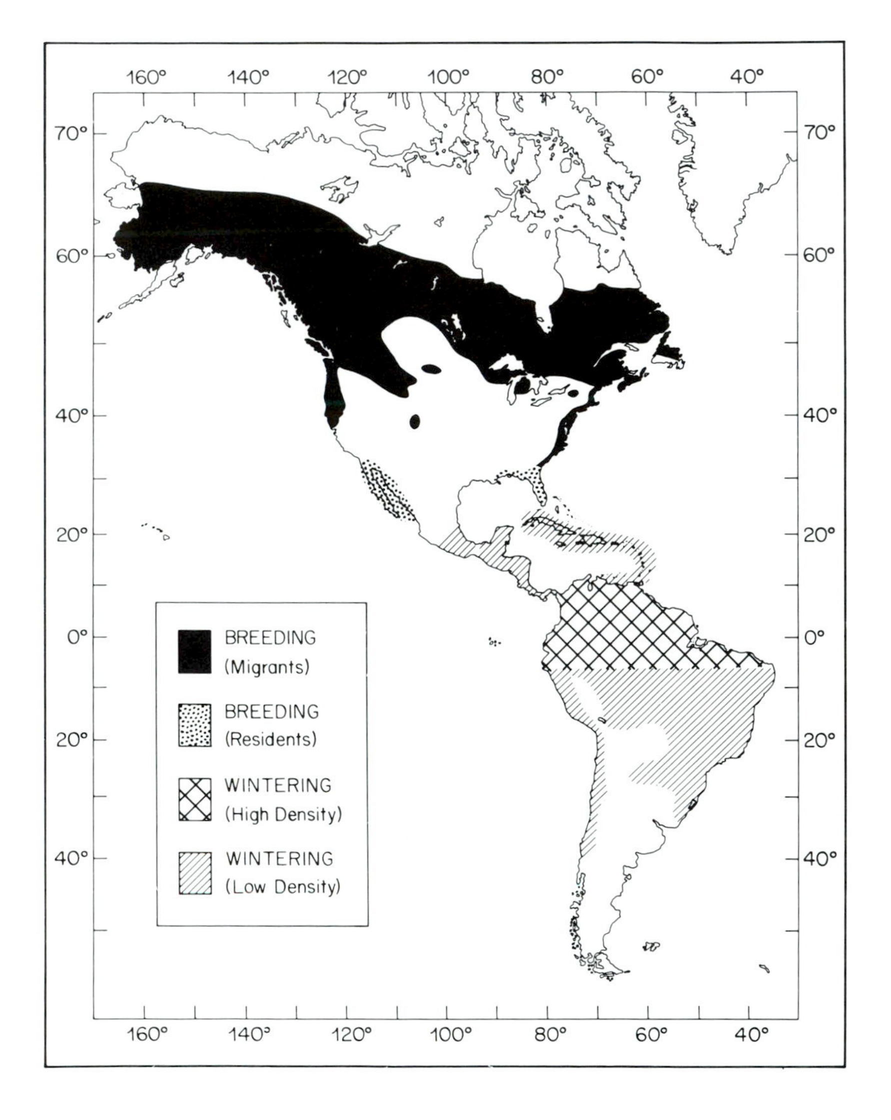

Figure 3.7. The breeding and wintering ranges of North American Ospreys (*P. h. carolinensis*). From Prevost (1983a) and Poole & Agler (1987).

United States: the lower 48

The US Osprey population is concentrated along the east coast, and apparently always has been. About 50% of all US Ospreys nest within foraging distance of the Atlantic seaboard or the Gulf of Mexico (Fig. 3.8). The vast, shallow expanse of Chesapeake Bay alone supports over 1500 breeding pairs, about 20% of the nation's total (Henny, 1983). Florida holds another 20%, and Maine about 12%, although some of these birds breed along inland lakes. Many eastern US Ospreys have taken advantage of safe nesting sites in coastal swamps or on predator-free islands, and large nesting colonies, occasionally numbering over 100 pairs, have formed in such locations (Fig. 3.9; Chapters 8 & 11). Eastern Ospreys have become increasingly dependent on artificial nest sites in recent decades. In many regions, 50%–75% of these birds now nest on poles, towers, or buoys (Fig. 3.10; Chapters 6 & 10). Where plentiful, such structures have concentrated breeders, often changing nesting distribution from that of earlier decades when only trees or island beaches were available to the birds (Reese, 1969).

During the 1950s and 1960s, pesticide contamination threatened many east coast Osprey populations. Egg viability fell drastically,

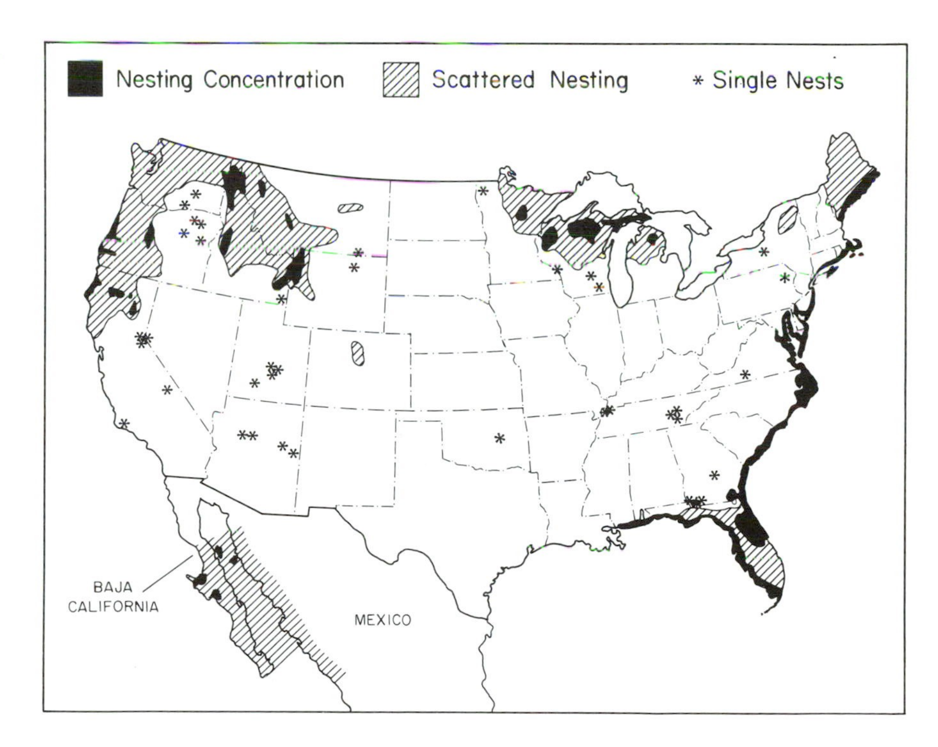

Figure 3.8. The distribution of Ospreys nesting in the United States and Mexico, about 1985. From Henny & Anderson (1979), Henny (1983), and unpublished sources.

depressing hatching rates and eventually breeding numbers as well (Chapters 9 & 11). Today, these populations are recovering. Reproductive success is high, breeding numbers are expanding rapidly, and pairs are adapting to an increasingly developed coastline (Chapter 11). Only northern Florida Bay is currently losing breeders, not from contaminants but from ecological changes that have made fish scarce, perhaps shifting Ospreys to the nearby Florida Keys where food and artificial nesting sites are plentiful (Ogden, 1977; Poole, 1982a; Kushlan & Bass, 1983).

A second but smaller US Osprey concentration breeds in northern Michigan, Wisconsin, and Minnesota, states that border the western Great Lakes (Fig. 3.8). These birds, isolated from others in the United States, are really a southern extension of the Canadian population. Very few pairs nest along the Great Lakes themselves. Most are scattered inland along swamps, rivers, and small lakes, their nests dotting the tops of tall conifers. Shallow inland lakes and reservoirs have concentrated about 30% of this population at a few key sites, with artificial nesting platforms helping in this regard. Although pre-1960 numbers in this region are not well known, midwestern Ospreys reproduced poorly during the era of reckless pesticide use, and diminished by about 50% (Postupalsky, 1977). Since the early 1970s, these birds have regained breeding strength and numbers. About 650 pairs now nest annually in these three

Figure 3.9. Small islands of mangrove trees, like those seen here in south Florida, provide favored nesting habitat for coastal Ospreys throughout the subtropics. Surrounding bay waters are generally shallow and productive. (Photo: M. Male.)

northern states (Postupalsky, 1983; Eckstein & Vanderschaegen, 1988; Henny, 1983). Continued growth in numbers can be expected.

A third US nesting population of about 1400 pairs lies scattered in a broad arc from the northern Rocky Mountains to the northern coast of California (Fig. 3.8). Most are fresh water Ospreys, dependent on

Figure 3.10. A typical Osprey nest site (an old channel marker) along the east coast of the United States. (Photo: A. Poole.)

lakes, rivers and, increasingly, reservoirs. Western reservoirs, generally shallow water floodings with abundant fish and dead tree nest sites, have allowed Ospreys to expand both range and numbers (Swenson, 1981). Twenty percent of California's and 50% of Oregon's Ospreys now nest near reservoirs, often in loose colonies (Henny, 1983). Significant breeding concentrations also occur at natural sites such as Yellowstone Lake in northwestern Wyoming and Flathead Lake in Montana (Swenson, 1979a; MacCarter & MacCarter, 1979). Surprisingly, few Ospreys breed along the US Pacific coast (Henny, 1983), probably because it is generally steep and rugged with few shallow inlets where fish are easily caught.

Overall, the western US population has been slowly expanding its range and numbers in recent decades, and shows promise of continuing this trend (Henny *et al.*, 1978b). Only in southern and central California, where persecution eliminated most Ospreys before 1920 (Henny, 1983), has this population lost ground.

Northwestern Mexico

An estimated 800–900 pairs of Ospreys nest along Mexico's Baja peninsula and in the nearby coastal states of Sonora and Sinaloa (Fig. 3.8; Henny & Anderson, 1979). This is desert country, so nests are confined to the coast, most built on rocky cliffs and pinnacles above the sea. Tall cacti are also favored nest sites here (perhaps the ultimate predator deterrent) and a few ground-nesting colonies have formed on coastal islands. One such colony, on narrow Great Whale Island in San Ignacio Lagoon, boasts over 100 pairs breeding along a four to five kilometer strip (B. Reitherman, unpublished). This is probably the largest, densest concentration of nesting Ospreys in the world today, a perfect example of how shallow, productive waters and safe, abundant nest sites combine to stimulate the growth of Osprey populations.

This Mexican population appears stable at present, despite low breeding rates (Judge, 1983), probably in part because the coastline is so inaccessible. As roads are developed, however, these birds could be threatened. Shooting and nest robbing, which eliminated Ospreys from accessible regions of northwestern Baja early in this century, are still threats in this impoverished country (Henny & Anderson, 1979).

Canada and Alaska

With few exceptions, Ospreys in Canada and Alaska are confined to the wide belt of coniferous forest that stretches from the Atlantic seaboard to the coasts of British Columbia and western Alaska (Fig. 3.7). The vast majority of these Ospreys breed along inland lakes and rivers, perhaps not surprising in a country where water is nearly as plentiful as land. As in the United States, coastal nesters are plentiful only along the Atlantic seaboard (Nova Scotia and New Brunswick) where shallow bays and coastal islands have encouraged small breeding colonies, a few near cities (Prevost, Bancroft, and Seymour, 1979; Stocek & Pearce, 1983; Greene & Freedman, 1986).

Aerial surveys flown over selected regions of Labrador, Quebec, Ontario and Saskatchewan found about 0.4 pairs / 100 km^2 near large rivers and about 1.4–3.4 pairs / 100 km^2 near lakes, far lower densities than are found in the boreal forests of Fennoscandia (Wetmore & Gillespie, 1976; Grier, Sindlar, & Evans, 1977; Bider & Bird, 1983; Scott & Houston, 1983). Most Canadian nests are in tall shoreline trees or on large boulders isolated by water; 40%–60% of the nests lack occupants, a high percentage for the species.

Two questions about Osprey distribution in Canada remain unanswered. First, it is not clear why this population fails to reach treeline. Several forested regions of northern Canada that I have canoed lack Ospreys. Fennoscandian Ospreys nest at much higher latitudes, right up to treeline, despite spring thaw dates that are just as late as northern Canada's (Saurola, 1986). Second, the Bald Eagle, another fish-eating raptor, does breed to treeline and has a much wider range than the Osprey in Canada. Where this eagle is plentiful, Canadian Ospreys are particularly sparse (Gerrard, Whitfield, & Maher, 1976). Could Bald Eagles be excluding Ospreys from suitable habitat in Canada? This is not the case in south Florida (Ogden, 1975), but perhaps ecological differences make the two species more competitive at northern latitudes.

Ospreys breed at low density along the southeast coast of Alaska (despite abundant islands) but have colonized no other portion of Alaska's coastline. Inland, recent aerial surveys by Jeff Hughes (1986) have located scattered pairs, except near Tetlin Lakes (63° N, 143° W) where a cluster of 20–30 pairs are breeding. Presumably this concentration reflects some ecological anomaly. Hughes suspects that spring thaws, which come earlier to Tetlin than to other Alaskan lakes, give Ospreys an early start on breeding and thus a better chance of raising young. Perhaps there are similar undiscovered 'hot spots' for Ospreys in Canada and Alaska.

3.5 Why do Ospreys not breed on their wintering grounds?
The previous sections have shown that Ospreys can reproduce successfully in a great variety of habitats. Yet we find three conspicuous and puzzling gaps in their breeding distribution: Africa, South America, and Indomalaysia – in short, the migratory Ospreys' primary wintering grounds. There are, to be sure, one or two isolated instances of Ospreys breeding in South Africa (Dean & Tarboton, 1983). There are also areas of slight overlap between breeders and migrants in the Caribbean, the Mediterranean, the Red Sea, and the Persian Gulf (Figs. 3.1 and 3.7). But in general, breeding Ospreys avoid the tropical latitudes that attract most migrants in winter. Australasian Ospreys, for example, have never moved north of the equator into Southeast Asia where migrants from the eastern USSR spend the winter (Fig. 3.5), and nonmigratory Palearctic Ospreys penetrate tropical latitudes only where wintering migrants are scarce or missing: in the Cape Verde Islands and along the southern Red Sea and the Arabian Peninsula (Fig. 3.1).

Yves Prevost (1982) spent over a year in Senegal, West Africa, studying Osprey wintering ecology and considering potential barriers to nesting in this region. He could find no ecological reason why Ospreys should not nest in Senegal or, by extension, elsewhere in the tropics. Osprey foraging efficiency was as high in Senegal as on most European breeding grounds. Likewise, climate is no barrier; Ospreys successfully raise young in tropical regions such as northern Australia and the southern Red Sea during months when average temperatures are considerably hotter than those experienced in most of Africa and South America. Nest site availability is no barrier either. Senegal and other key Osprey wintering sites possess abundant mangrove swamps, prime nesting habitat for resident Ospreys throughout the subtropics. Some ecologists have suggested that African Fish Eagles (*Haliaeetus vocifer*) might exclude Ospreys from portions of Africa (Newton, 1979; Brown, 1980). The two species are ecologically similar and the eagle does pirate fish from Ospreys. Yet these two raptors often coexist peacefully, even when sharing the same lake (Boshoff & Palmer, 1983). No other fish-eating bird of Africa or South America competes strongly with the Osprey in both marine and freshwater habitats. In short, most tropical habitats appear more than adequate for nesting Ospreys. Something else must prevent these birds from breeding on their preferred wintering grounds.

Breeding Ospreys might be expected to colonize their wintering grounds in either of two ways: as migrants lingering on to establish a nesting population or as nearby residents dispersing to fill vacant habitat. Northern migrants usually retain their northern breeding schedules while wintering in the tropics, however, so lingering migrants would be unlikely to breed (Immelmann, 1971). Adjacent resident populations, however, whose breeding schedules are well adjusted to tropical seasonality, would not be troubled by such a physiological barrier. Why have these residents failed to penetrate where migrant Ospreys winter at their greatest densities?

One answer is that migrants, where numerous, exclude resident breeders from their wintering grounds. Such purported competition between migrants and residents seems especially likely because winter, the appropriate breeding season for Ospreys in the tropics (see Fig. 6.1), is also the time when northern migrants reach their peak densities at tropical latitudes (Prevost, 1982). It is not clear what form such competition might take. Direct evidence of competition among and within species is notoriously elusive (MacArthur, 1972), especially among birds like Ospreys which share feeding sites. Prevost (1982) suggested that tropical breeders might be vulnerable to piracy by migrant Ospreys, yet Ospreys rarely steal fish from each other. Nor is it likely that migrants exhaust local food supplies; the presence of hunting Ospreys may actually help a neighboring bird locate fish (Chapter 5). Thus while the Osprey's biogeography strongly suggests that migrants exclude breeders, we lack convincing proof.

Knowing where the world's Ospreys breed, information gleaned from this chapter, we turn our attention to Ospreys outside the breeding season. The questions considered in the following chapter are: where do Ospreys winter, how do they migrate to their wintering grounds, and what is known about their wintering ecology?

4 MIGRATION AND WINTERING ECOLOGY

Hemisphere solidarity is new among statesmen, but not among the feathered navies of the skies.

Aldo Leopold (1949)

In June of 1972, the US Fish and Wildlife Service received the following letter, typical of hundreds that fill their files;

> Dear Gentlemen: . . . We wish to inform you that on the 27th of December at four in the afternoon around Veneral, River Yurumagui, in the municipality of Buenaventura, Valle de Cauca, Colombia, South America, a falcon was hunted which bore on its leg a band with the following number and inscription . . . This bird was in a perfect state of health and . . . it didn't seem to be very old . . . If you need other information on this animal you may address yourselves to . . . he who shot it with a caliber 16 rifle and ammunition, #4.

The 'falcon' this enthusiastic Colombian shot was actually an Osprey, born one year earlier along the Maryland (USA) coast and banded (ringed) the week before it fledged with a numbered aluminum leg band. That band was the key to this bird's identity, proof of its 4000 kilometer trip to South American wintering grounds. Had this Osprey lived to the ripe old age of 15, migrating each spring to breed near its natal site and returning each fall to winter in Colombia, it would have traveled over 100 000 kilometers during its lifetime. During an equally long life, other Ospreys – Canadian breeders that winter in Peru, for example, or Scottish Ospreys that winter in Cameroon, West Africa – might fly nearly twice that distance. Clearly the Osprey is a mobile creature, familiar with vast distances and a shifting complex of weather, prey, and habitat. We who travel such distances sleeping in jets have difficulty appreciating how extraordinary its journeys are.

This chapter examines the timing and geography of Osprey migration, concentrating on differences within and between populations. Such comparisons are possible only because the movements of several different populations have been carefully analyzed. Differential migration is of interest because, by showing how some populations respond to pressures that others can ignore, it sharpens our knowledge of how natural selection shapes the migratory movements of a species. North temperate Ospreys, for example, undertake long-distance migrations, while most subtropical Ospreys move only locally after breeding. Variation in migratory behavior thus makes the Osprey a particularly appropriate species in which to study the ecology and evolution of migration – why, that is, birds leave familiar breeding areas each year and undertake potentially hazardous journeys, and why some go farther than others.

4.1 Band recoveries and visual sightings

Compared with Ospreys at the nest, we know little about Ospreys during migration and on their wintering grounds. Without a nest site to hold them, individuals tend to scatter over remote regions and are often hard to find. Knowledge of Osprey migration, therefore, comes primarily from banded birds, although visual sightings at a few concentration points have provided some supporting data.

A band is of no help to a biologist until it has been 'recovered' – found on a dead or injured bird, with the number, location, and date of finding reported to a central banding office. Band 'returns', identifications of live birds (usually via color-bands), occur less often and mostly at breeding sites, so they contribute less to studies of migration. It takes dozens of recoveries to determine a population's migration route and wintering distribution, and recoveries are generally slow to accumulate.

Ospreys have been a particularly rewarding species to band, in part because they are easy to locate. Pairs often nest in accessible locations – low to the ground, in colonies, and near people – so banders have been encouraged to return to the same nesting areas year after year, building up large samples. (Most Ospreys are banded as nestlings, before they fly; banding adults is more difficult (section 7.1)). During the past 40–60 years, nearly 20 000 nestlings have been banded in the United States (most along the east coast) and nearly 14 000 in Finland and Sweden. It is these three countries where Osprey banding has been concentrated and thus these three populations whose migrations are best known.

The long-term efforts of a few Osprey banders is almost legendary. Sten Österlöf in Sweden, Pertti Saurola in Finland, and Jan Reese, Mitchell Byrd, Leroy Wilcox, and Joe Jacobs in the eastern United States together have accounted for nearly 40% of the Ospreys banded in the world. Jacobs climbed trees to band Osprey nestlings until he was well into his 70s and died of a heart attack one hot afternoon while trying to reach yet another nest. Österlöf fell from a high nest tree at nearly the same age and badly injured himself. Fortunately, banding Ospreys is rarely this risky and continues to attract new recruits.

Depending on location, 7%–11% of the Ospreys banded during the past 50 years have been recovered (Österlöf, 1977; Poole & Agler, 1987). Although shooting has accounted for 30%–40% of these recoveries, most have resulted from injuries, drownings, exhaustion or, in many cases, causes that were never known or specified by the person finding the bird.

Band recoveries have some potential biases, although they appear less critical for Ospreys than for most other species. Not only does the quantity and accuracy of information reported with recoveries vary but, as Österlöf (1977) has pointed out, language barriers or the simple cost of a postage stamp often prevent people in less developed countries from reporting a band. In addition, reporting rates vary with birds' ages and the timing of their migrations. Young birds, migrants, and individuals traveling during hunting seasons are more likely to be found or killed than are experienced and sedentary individuals (Newton, 1979; Poole & Agler, 1987). Hence, one cannot ignore geographical biases when analyzing band reports for a long-distance migrant like the Osprey.

Counting Ospreys and other migrating raptors at concentration points is an increasingly popular activity among birdwatchers, but such counts are often inaccurate. A simple change in weather conditions, for example, can scatter a stream of hawks passing along a coast or mountain ridge, or push the birds too high to see, making true counts elusive, even with the help of radar (Newton, 1979; Kerlinger & Gauthreaux, 1984). On wintering grounds, Ospreys have been studied and censused most often near human settlements, so in remote regions their numbers and distribution remain poorly known. Despite the inherent limitations of sightings and band recoveries, however, these two methods provide the only useful data we have to determine patterns of Osprey distribution outside of the breeding season. And compared with most other species of birds, these patterns are fairly clear, at least in broad outline.

4.2 Fall migration

Beginning in late summer each year, Ospreys breeding in north temperate regions, along with their newly fledged young, join hundreds of other birds in a massive exodus south to tropical wintering grounds (Fig. 4.1). Tracked from a satellite, and speeded up in time, this movement might appear as a broad, braided, south–flowing river, with some of the migration streams wider and deeper at various times than others. What accounts for this variation in the southward flow of Ospreys is examined below.

Dispersion, routes, and mode of migration

Fall recoveries of banded Ospreys suggest that these birds migrate south on a broad front. During September, for example, Ospreys from the northeastern United States scatter south and west of banding (nesting) locations (Fig. 4.2), typical movement for other US and Fennoscandian populations during autumn.

Although most fall recoveries (75%) are of juvenile birds (young-of-the-year that might be expected to wander on their first flight south), there are few indications that adults follow a more restricted route. Adult Swedish Ospreys, like their young, have been recovered during autumn in most of the countries of Western and central Europe (Österlöf, 1977). Adults have consistently outnumbered juveniles at one mountain hawk watch in the eastern United States, however, so there may be some routes that migrant juveniles avoid (Heintzelman, 1983).

Although migrating on a broad front, Ospreys do not ignore landforms that can help to guide and concentrate their movement. The species appears regularly along north–south mountain ridges in the eastern United States, for instance, and along the peninsulas of

Figure 4.1. Migrant Osprey in flight. These birds nearly always travel alone. (Photo: M. Male.)

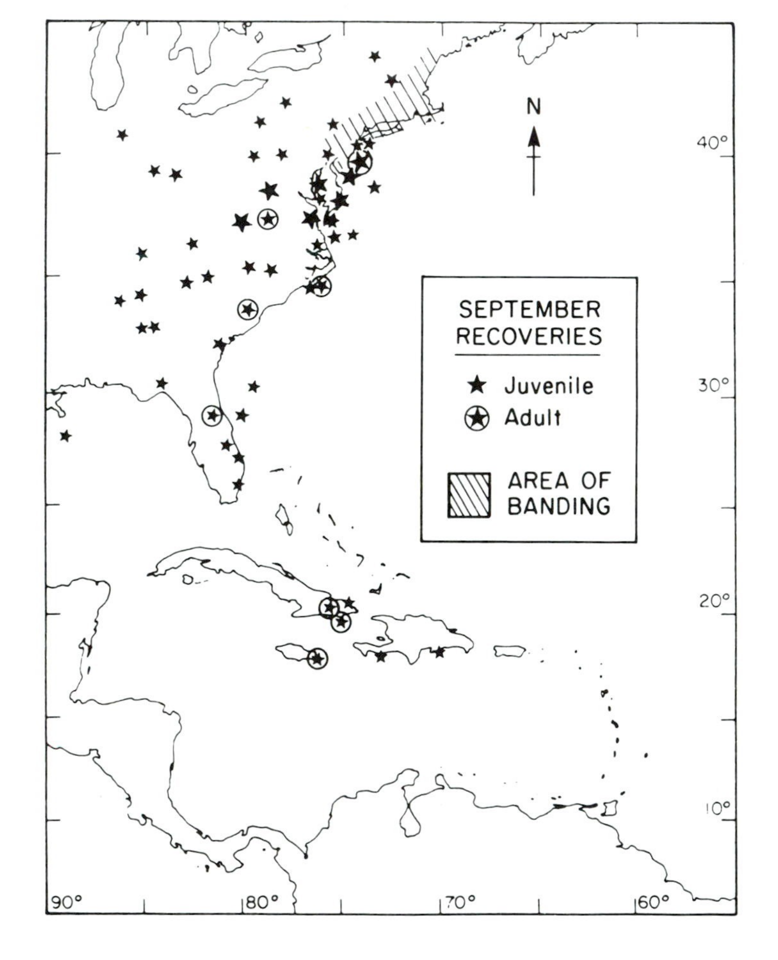

Figure 4.2. Sites where Ospreys banded in the northeastern United States were recovered during September, 1914–1984. Larger symbols indicate three recoveries within 50 kilometers of each other. Recoveries within 50 kilometers of the banding location have been excluded and three recoveries south of the map border are not shown. From Poole & Agler (1987).

Florida and Italy (Fig. 4.3). The latter finding suggests that both US and European populations may be shortening their migrations across the Caribbean and Mediterranean seas by funnelling along such routes (Henny & Van Velzen, 1972; Melotti & Spagnesi, 1979). Travel along the US Atlantic coastline appears more dense than at inland sites (Fig. 4.3), perhaps because most eastern US Ospreys originate from coastal colonies or because the coast offers abundant food. One golden day in early October, I watched nearly 800 Ospreys flying south along the coastal dunes of Cape May, New Jersey, an unusually dense concentration. Small, temporary groups of Ospreys are sometimes seen feeding and resting along rivers during fall and spring, but it is not clear if these migrants actually follow rivers to the same extent that they do peninsulas and coastlines.

Most diurnal birds of prey avoid overwater crossings of more than 15–20 kilometers because water never generates the updrafts and thermals that make for efficient soaring flight. Instead, most hawks and eagles prefer to circumvent the intervening body of water, thus concentrating along shorelines, land bridges, or at narrow water crossings (Newton, 1979; Kerlinger, 1985). Ospreys are an exception to this rule. During fall and spring, for example, they are seen far more regularly on the island of Malta in the central Mediterranean than crossing the Bosporus or the Strait of Gibraltar – narrow water gaps at either end of the Mediterranean where many other European raptors concentrate during migration (Beaman & Galea, 1974; Evans & Lathbury, 1973; Österlöf, 1977). Similarly, most fall and spring

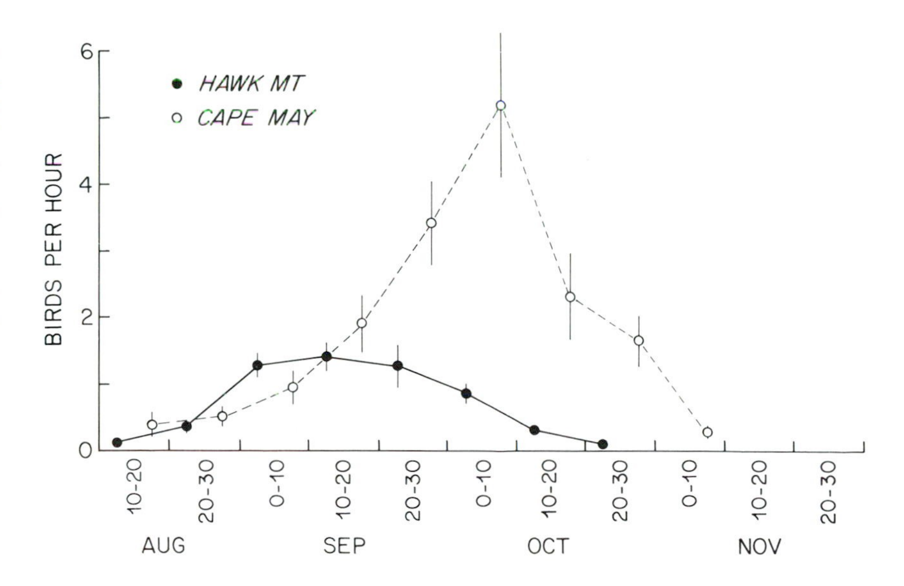

Figure 4.3. Fall sightings (1980–1984) of Ospreys at two hawk watch points in the eastern United States: an inland observation site at Hawk Mountain, Pennsylvania and a coastal site at Cape May, New Jersey. Data shown are the mean number of Ospreys counted per hour, vertical bar indicates ± one standard error. From Hawk Mountain Sanctuary and Cape May Bird Observatory (unpublished).

recoveries of US Ospreys in the Caribbean basin come from the large islands of the Greater Antilles: Cuba, Hispaniola, and Puerto Rico (Henny & Van Velzen, 1972; Poole & Agler, 1987). This suggests that Ospreys do not funnel south along the chains of smaller Caribbean islands (the Bahamas and Lesser Antilles) or along the Central American coast, but instead fly directly across the Caribbean to their South American wintering grounds, perhaps breaking up the journey by stopping in the Greater Antilles, a convenient halfway point.

Additional evidence that the Osprey is a capable overwater migrant comes from shipboard observers. In the western North Atlantic, Ospreys have been regularly seen flying south in fall, a few as far as 200 kilometers off the US coast and most well out of sight of land (Bent, 1937; Kerlinger, Cherry, & Powers, 1983). Generally these birds were strong fliers that did not seek ships to rest on, so there is no reason to think of them as lost or doomed. Such sightings suggest that some Ospreys from eastern North America may, like Peregrine Falcons, migrate offshore much of the time, returning sporadically to feed along the coast. Many shorebirds, some passerine birds, and some ducks and herons follow a high altitude route directly across the North Atlantic to the West Indies and South America, powered by extensive fat reserves (Williams, Ireland & Teal, 1977). It seems unlikely that Ospreys do this, however; they do not lay on much fat before migrating and would probably starve during the 70–80 hour trips if they did not stop to feed. Ospreys have appeared as vagrants on the islands of Hawaii and Bermuda, several thousand kilometers from their nearest breeding grounds, but most probably do not survive such long flights (Bradlee *et al.*, 1931; Stager, 1958; Munro, 1960).

Ospreys cross deserts as regularly as they do water. The Sahara presents a major ecological barrier to Palearctic migrants moving north or south (Fig. 4.4). Like many other Eurasian migrants, it seems that Ospreys cross the roughly 2000 kilometers of the Sahara in a single flight, a trip of 40–66 hours, provided that an individual flies continuously at flight speeds of 30–50 km per hour (Moreau, 1972; Curry-Lindahl, 1981). Moreau suggested that some Eurasian falcons make continuous, high-altitude flights over both the Mediterranean and the Sahara en route to sub-Saharan wintering grounds, but fall recoveries of Ospreys along the North African coast indicate that many stop there to rest and feed after crossing the Mediterranean, before attempting the Sahara (Figs. 4.4 and 4.9). Such migration in stages is probably the rule for Ospreys, even when covering shorter distances over more hospitable territory.

The Osprey's morphology and mode of flight facilitate direct passage over water and deserts. Many other birds of prey are predominantly soarers, restricted to the landforms that generate updrafts and thermals. While migrant Ospreys often do soar on updrafts when they find them, they are not exclusively 'light-air' migrants. Like other narrow-winged raptors (harriers, for example) with relatively high wing loading, they can fly actively and directly during migration, even traveling at dawn and dusk when thermals have died. Beaman & Galea (1974) regularly saw Ospreys arriving just after dawn on the island of Malta, so during longer water and desert crossings these birds undoubtedly travel at night. Over more hospitable territory, however, the Osprey is primarily a daytime migrant.

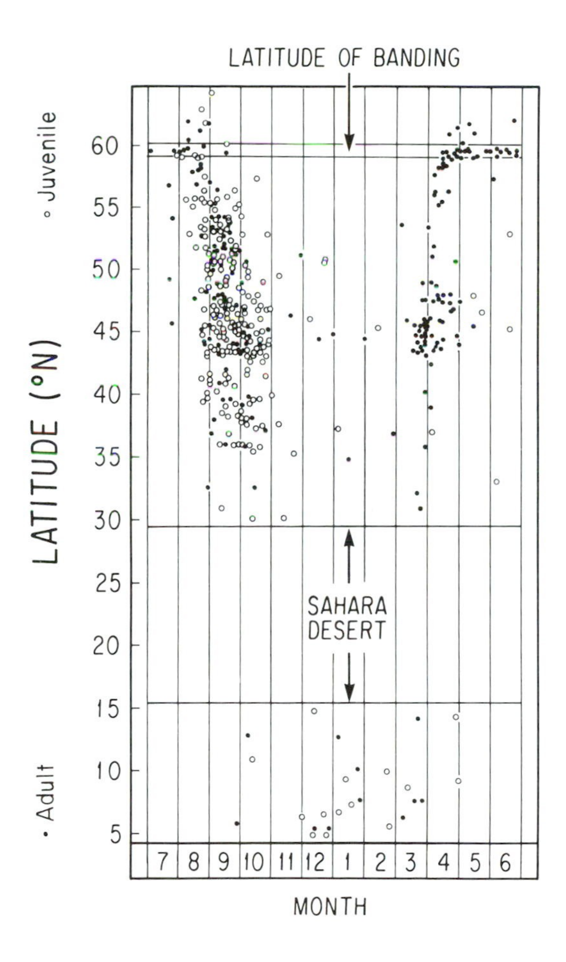

Figure 4.4. The north–south movement of adult and juvenile Ospreys during the course of the year, as shown by recoveries of birds banded in Sweden between 59°–60° N. Adults were three years of age or older, juveniles less than one year. Recoveries north of the banding area were birds that did not migrate directly south in fall or wandered north of their breeding grounds in spring. From Österlöf (1977).

Because Ospreys can migrate directly, breeding populations separated by longitude tend to maintain that separation during migration. Ospreys originating in Finland, for example, follow a more easterly route than Swedish birds, with the bulk of the Finnish population passing through eastern Europe, the western USSR, and the Greek and Italian peninsulas in fall (Österlöf, 1977). Most Swedish Ospreys filter through western Europe, although recoveries from both populations do overlap to some extent in Italy (Melotti & Spagnesi, 1979). Of the few Scottish Ospreys recovered on passage, all were found in Portugal, western France, or Spain, well west of most Swedish birds (Dennis, 1984). Eastern, midwestern, and western US breeding populations also remain largely separate during fall migration (Fig. 4.5).

Timing of migration

Data gathered from hawk watch points, from band recoveries, and from scattered observations at nests all indicate that Osprey migration is well underway at north temperate latitudes by the last two weeks of August (Figs. 4.3, 4.4, and 4.5). During September and continuing into early October, numbers peak at watch points in both

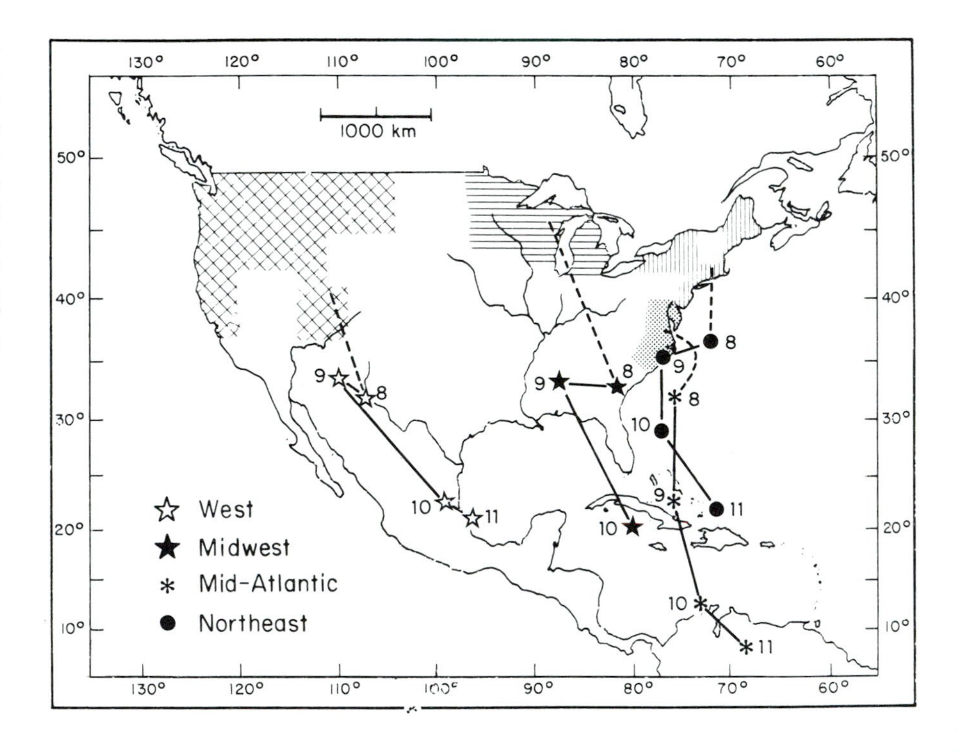

Figure 4.5. Mean monthly coordinates of recovery during fall migration for Ospreys (all ages) originating from four separate US breeding populations. Numbers indicate month of recovery; shaded areas show regions of origin (banding). Dashed lines are the initial legs of migration. From Poole & Agler (1987).

the United States and Europe (Fig. 4.3; Österlöf, 1977). During this same period, however, a few Ospreys are also recovered in Caribbean and Mediterranean countries, more than 20 degrees south of breeding sites. By early November, numbers drop off at watch points in north temperate regions and recoveries strengthen at more southerly latitudes. On Senegambian and Mexican wintering grounds, a few new arrivals appear as early as late August, with most birds arriving during October (Fig. 4.8; Prevost, 1982). For Ospreys breeding at north temperate latitudes, therefore, migration is spread over a two to three month period, with considerable variation in when individuals reach their wintering grounds.

Two related factors influence the timing of fall migration: the date individuals depart their breeding grounds and the speed at which they travel once migration is underway. Within populations, pairs late in hatching young are generally the last to leave the nest, while nonbreeders usually leave several weeks ahead of pairs with young; failed breeders often stay as late as successful pairs (Poole, 1984). There is no convincing evidence that one sex migrates ahead of the other, despite Moll's (1962) suggestion that males go first. Fledged young, however, may linger at nest sites after their parents have gone and are usually the last to leave (section 6.6).

Because Ospreys breed considerably earlier at lower latitudes and altitudes than at higher ones (Fig. 6.1), one might expect significant differences in nest departure dates depending on geography. There are few reliable field observations to permit such comparisons, although band recoveries help. During August, for example, US Ospreys from northeastern (41°–45° N) and mid-Atlantic (35°–38° N) states move equal distances south from their respective breeding grounds (Fig. 4.6), and a roughly similar percentage of each population is recovered near nest (banding) sites. This suggests that each population starts migration at about the same time. During September, October, and November, however, mid-Atlantic Ospreys move south faster than northeastern Ospreys, arriving near winter quarters well before these others (Fig. 4.6). The farther south that migrant Ospreys breed in the northern hemisphere, therefore, the faster they tend to complete fall migration.

Dividing migrants into adults and juveniles (Figs. 4.6 and 4.7) helps to show why southern US populations migrate faster than northern ones. Adults from the two populations move south at about the same speed. Likewise, mid-Atlantic young travel south at roughly the same rate as adults, but not because they follow their

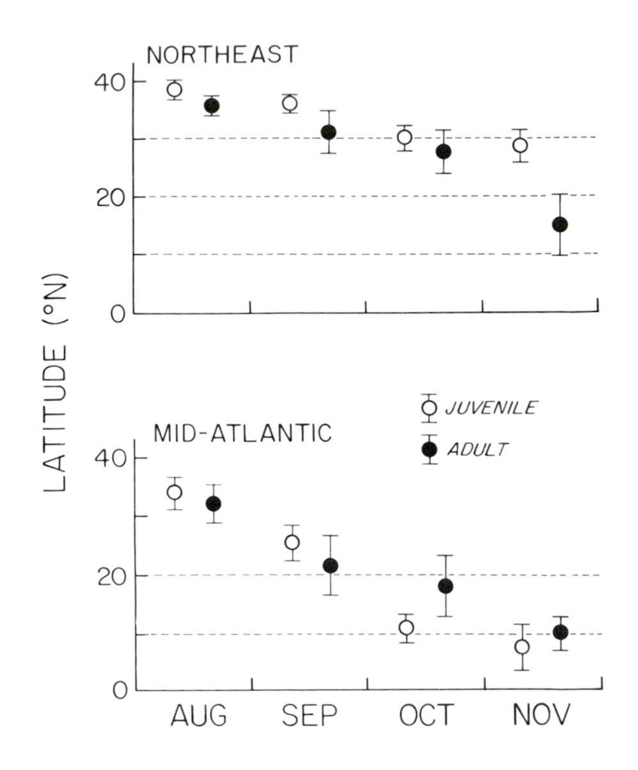

Figure 4.6. Mean and standard error of monthly latitudes of recovery during fall migration for adult (3+ years) and juvenile (<1 year) Ospreys banded in northeastern and mid-Atlantic states, USA, 1914–1984. (Poole & Agler, 1987).

Figure 4.7. An adult Osprey with its newly fledged young (speckled plumage) a few weeks prior to starting migration. (Photo: M. Male.)

parents. Like all Ospreys, these young migrate alone. Juveniles from northeastern states, however, lag increasingly behind their parents and far behind mid-Atlantic young, much farther than can be accounted for by the short distance between these two populations. Similarly, Österlöf (1977) found that young from Fennoscandian nests, well north of those in the United States, also lag behind their parents during fall migration.

Why should nesting latitude affect the speed at which juvenile Ospreys, but not adults, migrate south? One answer may be that Ospreys lay earlier, and thus fledge their young sooner, the farther south they breed at north temperate latitudes (Fig. 6.1). Young from the mid-Atlantic states, for instance, fledge two to four weeks earlier than most young from the northeastern United States, giving them extra time to gain weight and strength, as well as flight and foraging skills, before heading off into unfamiliar territory. Slowed juvenile migration thus seems more a function of fledging date than of any lack of competence inherent in being a juvenile. Indeed the ability of mid-Atlantic juveniles to complete their first migration as quickly as experienced adults shows clearly that familiarity with a migration route is not a critical advantage for an Osprey. This, in turn, suggests that inherited navigational directions, not learned landmarks, guide migrant Ospreys (Gwinner, 1986).

4.3 Winter quarters and wintering ecology

Timing: arrivals and departures

Observers from tropical regions report that most Ospreys reach winter quarters by late November and remain there, showing only local movement, until return migration begins in late February or early March (Prevost, 1982; Boshoff & Palmer, 1983; Fig. 4.8). Band recoveries support these observations. During December, January, and February, monthly shifts in the mean longitude and latitude of recovery for US Ospreys are negligible (Poole & Agler, 1987). During November and March, by contrast, Ospreys are recovered significantly farther north than in winter (Figs. 4.4 and 4.5), showing that some migration is underway during these two months. Thus 'winter' as used here is not an arbitrarily chosen period but rather corresponds to a relatively sedentary period in the yearly cycle of migratory Ospreys.

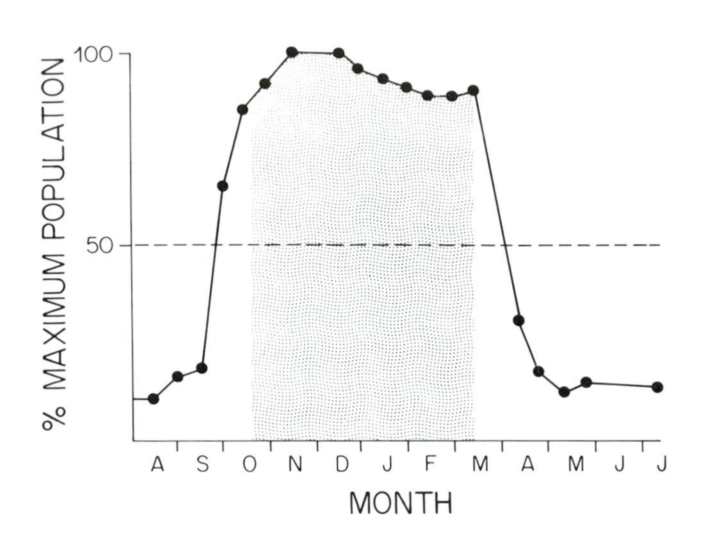

Figure 4.8. Monthly changes in the number of Ospreys censused on a large reservoir in southern Mexico (Oaxaca), 1981–1983. These were migrant birds, probably from the western United States; the maximum number seen here was 80. Shaded area shows period of peak density. From Barradas (1984).

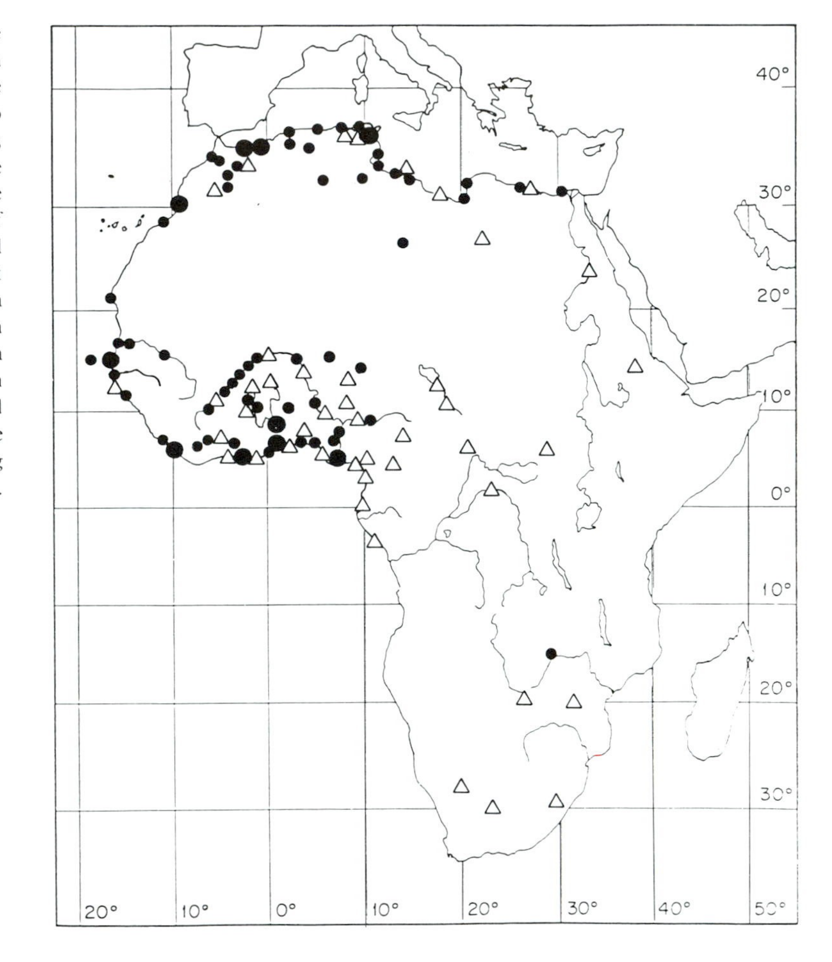

Figure 4.9. Locations where Ospreys banded in Fennoscandia were recovered in Africa, prior to 1975. Swedish Ospreys (●); Finnish Ospreys (△); large dots indicate three recoveries within 50 km of each other. Recoveries in all months are included but most recoveries were in winter, except in North Africa, where most were in fall and spring. From Österlöf (1977) and unpublished data of the South African Ringing Group.

Distribution

Although north temperate Ospreys winter across a wide belt of latitudes, most migrate to the tropics. Winter recoveries of Finnish and Swedish Ospreys are concentrated in West Africa, 4500–6000 kilometers south of their breeding sites (Fig. 4.9). Here, in the arc of coastal countries from Senegal south to Cameroon, Ospreys are supported by a highly productive network of interlacing ecosystems: rain forest, coastal mangrove forest, and broad, seasonally flooded rivers (Prevost, 1982). In central, eastern, and southern Africa, Osprey wintering density is much lower (Fig. 4.9), perhaps because lakes and rivers are less productive in those regions, because the populations traditionally migrating there have been reduced (Österlöf, 1965) or, in the case of South Africa, because it is so far from European breeding sites. Small concentrations of Ospreys do winter

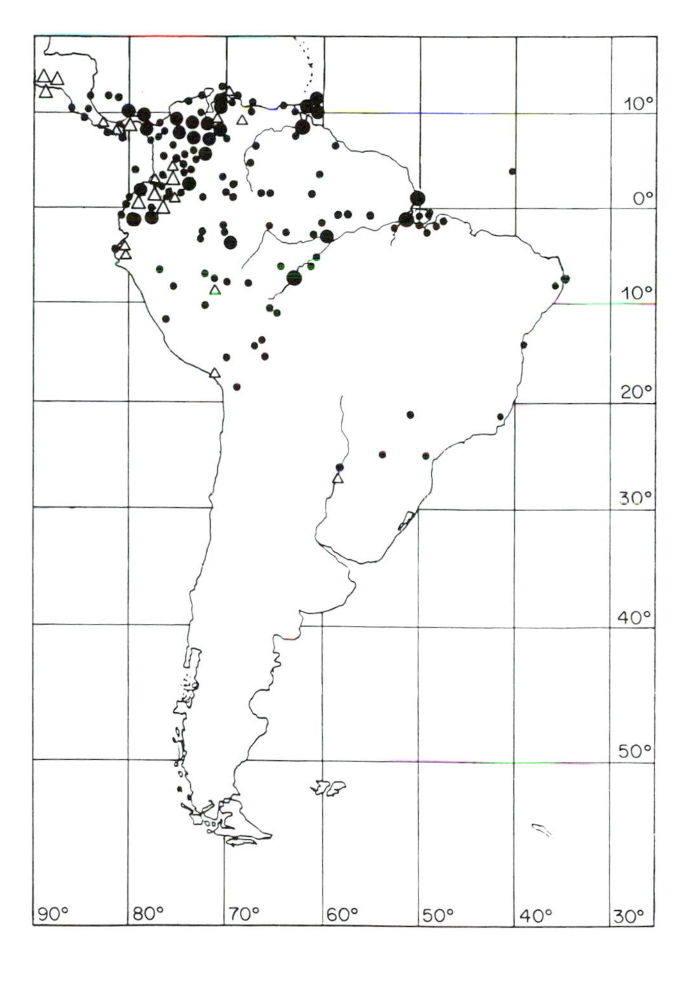

Figure 4.10. Locations south of 15°N where US Ospreys were recovered, 1914–1984, mostly during winter. Ospreys from east coast breeding populations (●); Ospreys from midwestern and western populations (△). Large dots and triangles indicate three recoveries within 50 km of each other. From Poole & Agler (1987).

in South Africa, however, so band recoveries may underestimate true numbers there (Boshoff & Palmer, 1983).

North American Ospreys also seek a tropical climate in winter. Band recoveries indicate that most US Ospreys winter in southern Central America and in northern South America (Figs. 4.10 and 4.11), at latitudes similar to those favored by European Ospreys wintering in West Africa. But because US Ospreys breed 15–20 degrees south of Fennoscandian Ospreys, they have a shorter migration, about 1000–2000 kilometers less each way. Many US Ospreys do migrate south of the equator, although few have been recovered south of 20° S and none south of 30° S (Fig. 4.10). Nevertheless Ospreys are rare but regular winter visitors to Chile (as far south as 40° S) and to southern Uruguay (35° S), so here again sightings provide information that band recoveries have yet to reveal (Schlatter & Morales, 1980; Blake, 1977).

Populations separated by longitude on their breeding grounds tend to show similar separation on their wintering grounds, although

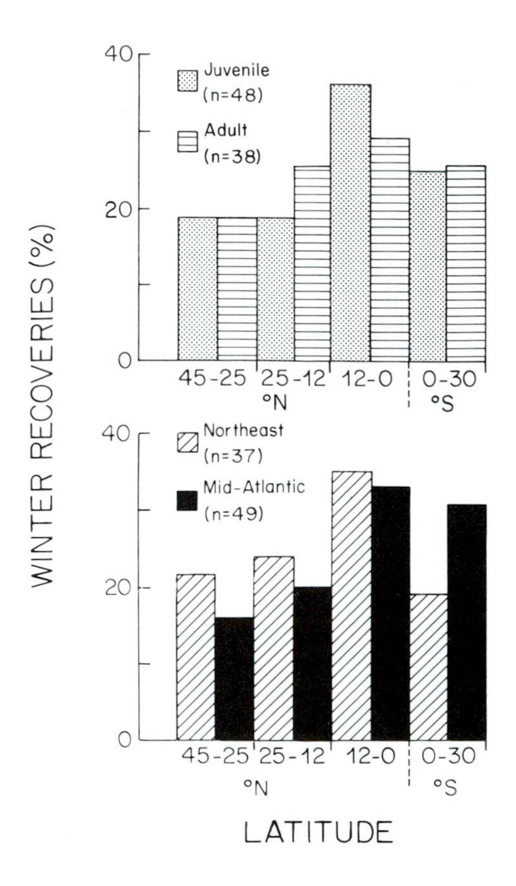

Figure 4.11. Latitudinal distribution of eastern US Ospreys in winter (December–February), as shown by band recoveries. Lower graph compares the distribution of birds (all ages) from northeastern and mid-Atlantic states; upper graph compares the distribution of juveniles (<1 year) and adults (2+ years) from both these areas combined. Although mid-Atlantic Ospreys seem more likely to winter south of the equator than northeastern Ospreys, the distributions of these two groups are statistically similar, as are those of the adults and juveniles. (Poole & Agler, 1987).

there is some overlap. Thus most US Ospreys wintering in eastern Brazil come from nests along the Atlantic seaboard, not the western United States, and most recoveries from eastern and central Africa are of Finnish, not Swedish, origin (Österlöf, 1977; Poole & Agler, 1987).

In addition, some individuals from both New and Old World populations winter much farther south than others, much farther than seems necessary just to survive. Fennoscandian Ospreys wintering in South Africa, for example, are 3300–4400 kilometers south of their counterparts in Senegal. Similarly, US Ospreys in southern Brazil have traveled nearly twice as far as those that winter in northern Colombia or Venezuela.

Who are these Ospreys that go so far? Differential migration is often attributed to age in other species of birds. Young, subordinate, or late individuals, it is thought, must travel longer distances than others to find unoccupied territory (Newton, 1979; Ketterson & Nolan, 1983). Yet neither age nor migration schedules seem to influence where Ospreys winter. Recoveries suggest that juvenile Ospreys winter at similar latitudes, on average, as their parents (Fig. 4.11, top; Österlöf, 1977). In addition, there is no convincing evidence that Ospreys engage in 'leap-frog' migration, whereby later migrants pass through areas occupied by earlier arrivals. In the United States, for example, northeastern and mid-Atlantic populations share roughly similar latitudes in winter, though the latter migrate well ahead of the former (Figs. 4.6 and 4.11, bottom). Likewise, most Finnish Ospreys breed north of those in Sweden but latitude does not separate these two groups on their wintering grounds (Österlöf, 1977). Thus how far an Osprey withdraws from its breeding ground is based neither on age nor breeding location, apparently, but on factors we do not yet understand.

A few Ospreys linger on in the United States and Europe during winter and never reach the tropics (Fig. 4.11; Österlöf, 1977). Most of these individuals are injured or sick, however, and probably cannot complete a full migration. Many undoubtedly die before spring. Ospreys wintering slightly farther south in northern Central America, the northern Caribbean, and along the southern shore of the Mediterranean probably are viable wintering populations, although their low density suggests that these regions may not be optimal habitat for the species. The southern coasts of Spain, Israel, Florida, or Louisiana might seem adequate habitats for wintering Ospreys, but the benefits of migrating an extra 1000–2000 kilometers south to fertile tropical rivers and estuaries apparently outweigh the costs.

Outside Africa and Latin America, Osprey wintering sites are known only imperfectly. Some Ospreys winter in India and Southeast Asia, scattered along coasts, inland lakes, and rivers (Smythies, 1960; Ali & Ripley, 1968; Medway & Wells, 1976; Fischer, 1983). Assuming rough longitudinal fidelity to nesting grounds, Ospreys wintering in Indomalaysia probably come from the eastern and central USSR. Ospreys also winter sparsely in Sri Lanka and on smaller archipelagoes in the Indian Ocean (the Laccadive, Andaman, and Maldive islands), further proof of this hawk's ability to cross open water. In Japan, northern Ospreys are thought to migrate south to Kyushu, where breeders are resident year-round, perhaps continuing south through the Ryuku Islands to Taiwan and the Philippines.

Ecology

Most of what is known about the ecology of wintering Ospreys stems from Yves Prevost's (1982) two-year study in coastal Senegambia, West Africa, a region where European Ospreys concentrate in winter. No one before Prevost had tackled the subject. His expedition focused on distribution, diet, habitat choice, social behavior, and foraging efficiency. Driving the beaches of Senegal, Prevost and co-workers, many of them local Wolof tribesmen, became experts at trapping Ospreys at feeding perches, using elastic powered snares. Hundreds were caught, marked, and released, providing data on weight, molt, and subsequent movements. Since many of these Ospreys had been banded as nestlings, their European origins were also clarified.

Like many other birds, Ospreys are remarkably faithful to wintering sites. Six of nine Ospreys marked at the mouth of the Senegal River, for example, were sighted there again the following year, after migrating to Europe and back. During a single winter, most stayed within 10 kilometers of their banding site. One marked bird was killed by a Crocodile (*Crocodylus niloticus*) 50 kilometers from its release point, but this was an unusually long distance for an Osprey of this region to travel in winter. (Presumably this Osprey was recovered because the Crocodile found it unappetizing, not because Prevost snatched it from the jaws of the hungry reptile.) During summer, however, when only immature birds remained in

Table 4.1 *Estimates of the population density and foraging efficiency of Ospreys wintering in key habitats in Senegambia, West Africa.*

Habitat	Density (no. of birds per km)	Foraging efficiency (kcal per min hunting)
Estuary[a]	3.3–0.9	5.2–9.7
Mangrove	0.1–0.4[b]	2.8–5.3
Beach	0.3–0.6	<3.2–10.0
Inland rivers	0.2	—

[a] River deltas, tidal mudflats, coastal ponds: shoreline and flats surrounding estuary.
[b] no. per km².
Based on surveys by Prevost (1982).

Senegal, Prevost found that Osprey numbers were not reduced everywhere in the same proportions. Immatures could have been moving to preferred habitat after the departure of adults. Likewise, J.E. Botero (personal communication) has suggested that most Ospreys wintering in the coastal marshes of Colombia move locally following seasonal shifts in water levels, salinity, and fish populations.

Senegambian Ospreys feed and rest in greatest numbers along coastal estuaries, especially shallow, open, tidal flats at river mouths (Table 4.1). Ospreys wintering in South Africa and in Panama show similar preferences (Boshoff & Palmer, 1983; P. Spitzer, unpublished). Estuaries are highly productive habitats where fish thrive. In the Senegal River delta alone, with a surface area of about 146 000 hectares, at least 30 000 tons of fish are eaten annually by vast flocks of fish-eating birds, and local fishermen take an equal or larger share (Curry-Lindahl, 1981; Prevost, 1982). For Ospreys, these estuarine food webs are particularly attractive because they are exploited most efficiently by schooling fish like mullet – abundant, easy prey in shallow water.

Ospreys wintering along estuaries often congregate in loose flocks. Prevost counted as many as 20–25 perched within 100 meters of each other and, in one instance, five perched on a beached log only five meters long – lined up like swallows on a wire. Such sociality may have been due in part to the lack of suitable feeding and resting perches in this open habitat, but hunting birds flocked too. Ospreys hovering and diving over schools of fish often stimulated others to join them. There appears to be a strong social component to all aspects of the Osprey's winter life in these open estuaries. Similar social behavior has been observed among Ospreys wintering in South African estuaries and along the Pacific coast of Panama (Boshoff & Palmer, 1983; P. Spitzer, unpublished).

Where mangrove forests impinge on the open estuaries, however, wintering Ospreys disperse and forage quite differently. In Senegal, at higher tides, Ospreys perch quietly on low mangroves overlooking the edges of creeks and pools, waiting for fish to swim within sight (Prevost, 1982). Foraging efficiency is low here (Table 4.1) and the birds are evenly dispersed, rarely within sight of each other, suggesting that spacing may be socially determined. Along the wooded shores of a Mexican reservoir, the pattern was similar (Barradas, 1984). The presence of suitable hunting perches, therefore, can change the way Ospreys forage and disperse. The ease of perching and waiting for fish apparently extends the time in which an Osprey can efficiently scan an area, making feeding territories worth defending.

Along exposed beaches in Senegal, Prevost found that Ospreys were distributed randomly at low density, not concentrated as they were along estuaries and not spaced evenly as along mangrove creeks. Beach perches were plentiful but most were near the ground, so at night the birds commuted two to three kilometers inland to roost in taller palm groves, avoiding potential predators such as Common Jackals (*Canis aureus*) and Spotted Hyenas (*Crocuta crocuta*). Foraging efficiency was low here (Table 4.1), except when schools of flying fish were drifted close to shore by strong winds.

Inland, few Ospreys wintered along the Senegal River; four times as many were found at the mouth of this river as upstream (Table 4.1). According to band recoveries, however, many Ospreys do winter inland along the Niger and Volta Rivers, south of Senegal, as well as along the vast network of rain forest rivers in South America (Figs. 4.9 and 4.10; Poole & Agler, 1987). Such distributional differences probably reflect differences in prey availability. Goulding (1980), for example, has described the extraordinary diversity of fruit and seed eating fish that live in the shallow flooded forests of the Amazon River basin. Ospreys wintering there probably depend on such fish, although this has yet to be studied.

Table 4.2. *Percentage of monthly band recoveries, grouped by region and age, made within 100 kilometers of various US breeding grounds during March–June 1914–1984. An Osprey's breeding ground was assumed to be near the banding location (Chapter 8). 3+ and 2 refer to Ospreys 3+ and 2 years old, respectively. East refers to mid-Atlantic and northeast recoveries combined.*

	Percentage recovered near breeding grounds			
	Northeast 3+	Mid-Atlantic 3+	Midwest 3+	East 2
March	9	66	0	—
April	66	91	25	36
May	85	81	100	39
June	83	87	100	67

From Poole & Agler (1987).

4.4 Spring migration

Timing

Prevost (1982) noted Ospreys leaving Senegal in early March, presumably headed north to Palearctic breeding grounds. Departures were gradual, but by late March numbers had decreased to less than half their winter levels. Temporary concentrations seen along Senegal's coast throughout March suggested that more southerly wintering populations were moving through behind the departing migrants. Sightings in Panama and Mexico indicate roughly similar timing for North American migrants in spring (Fig. 4.8; Wetmore, 1965).

Given these departure dates, sightings of Ospreys arriving at nest sites indicate a quick passage north for these birds. Most Scottish and Fennoscandian Ospreys arrive at nests during early and mid-April, respectively, about one to one and a half months after they leave West Africa (Fig. 6.1). Band recoveries confirm this speedy passage north. Even with juveniles excluded from the comparison, equivalent distances are traveled two to three times faster than during fall migration. Pressures to breed early are obviously great; early nesting Ospreys produce the most surviving young (Chapter 7).

Two-year-olds migrate more slowly than adults (Table 4.2). Such delayed migration, typical of young, nonbreeding birds, suggests that the constraints of breeding really do set migration schedules.

Because two-year-olds do not breed, they can afford to leave late or to migrate slowly.

One-year-old Ospreys are at least two years away from first breeding and so they gain no advantage by returning north their first spring. After their first migration south, juveniles remain in winter quarters for another 18–20 months until they return to breeding grounds as two-year-olds (Henny & Van Velzen, 1972; Österlöf, 1977; Prevost, 1982). Except in Australasia, therefore, Ospreys lingering south of 25° N during summer are presumably younger than two years.

Routes

Band recoveries and sightings suggest that Ospreys choose the same routes and modes of migration in spring as in fall: direct, broad front, in stages, and crossing seas and deserts where necessary (Beaman & Galea, 1974, Österlöf, 1977; Poole & Agler, 1987). In southern Israel, for example, where the spring passage of migrant raptors around the eastern end of the Mediterranean is the largest in the world, the Osprey is rare compared to most other raptors, reflecting its direct passage over the Mediterranean (Christensen *et al.*, 1981). I spent ten days in Israel in late March, 1987, and saw only one migrant Osprey among tens of thousands of other birds of prey – a lone male that flew steadily through desert heat up the Dead Sea Valley, enroute, no doubt, to some thawing, fog-bound lake in Finland or the USSR.

4.5 Resident Ospreys: short-distance movement

Ospreys from nonmigratory populations, most of which breed at subtropical latitudes (Chapter 3), are often called 'residents'. This term is misleading, however, because some residents do leave their nesting grounds during the nonbreeding season, although their movement is local compared with the long-distance migrations examined above. Some Australian Ospreys, for example, wander a few hundred kilometers inland during the nonbreeding season (section 3.2), although most remain year around at the coast. On Corsica, it is nonbreeders and young-of-the-year that tend to leave each fall and winter, perhaps for North Africa, while breeding adults can be found at nest sites in all months (Bouvet & Thibault, 1980). Ogden (1977) likewise reported that about half the Ospreys nesting in south Florida remained near colonies in the off-season and that most were adults. Band recoveries showed that subadults were most likely to leave, traveling north up the Florida peninsula, but seldom more than 100 kilometers. This movement included both migrants (individuals returning to their natal area) as well as dispersers (individuals that settled elsewhere), but recoveries were too few to establish any clear dispersal patterns. In northern Florida, where resident populations border on migratory ones, most Ospreys leave nests during fall and early winter, apparently dispersing locally, although a few inland breeders remain behind (T. Edwards & J. Reinman, personal communication).

Why do resident Ospreys undertake these local movements, and why are subadults most likely to leave? Assuming each population breeds at the most favorable time for its location, food availability probably dwindles during the nonbreeding season. With food scarce, experienced adults presumably have an advantage over younger birds, who can find more food by wandering. In addition, breeding adults invest much time in finding, building, and defending a nest. Remaining nearby during the nonbreeding season may guarantee access to that nest when breeding resumes. These are guesses, of course. Until someone actually observes nest defense and foraging success among nonmigratory Ospreys, it will be difficult to know why some individuals are more 'resident' than others.

Migration is hazardous. Undoubtedly, all Ospreys would be residents if they could get away with it. Yet one has only to look out a winter window to see why Ospreys leave most northern regions each fall. The birds themselves are probably not affected directly by the cold. Captive Ospreys held outdoors but fed regularly easily survive winter in the northeastern United States. Fish, however, are cold-blooded and thus sensitive to changes in water temperature. When temperatures drop, most avoid the colder shallows and surface waters where Ospreys can reach them. Thus even where northern lakes and bays do not freeze in winter, Ospreys attempting to winter over would probably starve.

If lack of fish is the ultimate reason for Osprey migration, one should be able to predict which populations will migrate by checking local winter temperatures. Major boundaries between migrant and resident Ospreys are found in the southern United States roughly in northern Florida and southern California, and in Europe along the northern coast of the Mediterranean (Figs. 3.1 and 3.7). Sure enough, a comparison of winter temperatures north and south of these boundaries shows that winter freezes are regular only where migrants breed (Appendix 2). Resident populations are rarely exposed to such low temperatures. Of course, temperatures are often warm and food still plentiful when northern migrants leave their breeding grounds, so hunger is not the immediate stimulus for departure. Rather it is an Osprey's internal clock, subtle changes in the flow of hormones, that no doubt generate the necessary restlessness (Gwinner, 1986). In this way, selection ensures that Ospreys vacate breeding areas well before their food becomes scarce, just as they vacate warm, productive tropical regions each spring when they return north to breed.

In summary, Osprey populations are migratory in regions where winter temperatures regularly go below freezing, reducing the availability of their cold-blooded prey. This, and the fact that Ospreys leave earlier and withdraw farther from breeding grounds than raptors dependent on hardy, warm-blooded prey, indicate that migration in this species is ultimately (even if not immediately) controlled by food supply.

Clearly, each Osprey's success depends on its ability to find food in a variety of habitats. But what, specifically, is the Osprey's food supply? Do Fish Hawks eat only fish? How do they find and catch their food? What accounts for variation in hunting efficiency among individuals and populations? How efficient does a hunting Osprey need to be before it begins to lose weight? These are the key questions addressed in the following chapter.

5 Diet and Foraging Ecology

I will provide thee of a princely Osprey,
That, as he flieth over fish in pools,
The fish shall turn their glistering bellies up,
And thou shall take thy liberal choice of all.

(Peele, 1594, *Battle of Alcazar*, act 1, sc. 1)

Ospreys may lack the sheer mass and power of eagles or the dazzling aerial agility of some falcons, but their hunts are always exciting. Wolof tribesmen living along the beaches of Senegal chant of the Osprey's hunting skills as they paddle canoes through Atlantic surf to tend their nets. Skilled fishermen themselves, they share fishing grounds with Ospreys born thousands of kilometers away along Swedish lakes and Scottish lochs. These people clearly admire the prowess of a plunging fish hawk. It is an everyday sight for them, woven into the fabric of their culture. Yet one need not be a fisherman to appreciate such skill. There is tension and drama to be found here: tension in the Osprey's slow careful progress over sunlit waters, hovering, moving on, hovering again, eyes alert to movement below; drama in its steep sudden plunge, feet thrown forward and wings tucked back, a shower of spray; and drama in its strong, nimble lift from the water with a glint of silver fish struggling in its talons.

Like lions stalking prey on the open grasslands of East Africa, fishing Ospreys are conspicuous, spectacular. Their visibility is both pleasing and helpful to the student of this species. Indeed, diet and foraging behavior are among the better researched aspects of Osprey ecology. Although most studies have recorded only dive success or prey remains found at nests, three studies have gone beyond this to provide synthesis and insight. Yves Prevost (1977, 1982) watched thousands of Ospreys hunt in Nova Scotia, Europe, and West Africa. His work refined the study of Osprey hunting efficiency and pieced together the first complete energy budgets for the species. Jon Swenson (1979b), reviewing various studies of Osprey hunting success in different locales, showed ecologically why these birds caught some fish more easily than others. Borrowing from these and other studies, this chapter provides a glimpse of what Ospreys eat in various regions, how they catch their prey, and how efficiently individuals in different populations meet their own and their family's daily energy needs. We also look briefly at the Osprey's morphology: how its body is adapted for a life of catching and eating fish.

Consider the problems an Osprey confronts when hunting. It must first decide where to hunt and how to get there, relying on memory or an ability to find productive new sites. It must also take into account such variables as weather, tides, and fish migrations. Once it locates fish, an Osprey must pick out a vulnerable individual of the appropriate size, stalk it carefully, and then dive accurately enough to snag the slippery, elusive form under water.

Ospreys dive feet first. They are buoyant and, despite long legs, can penetrate only about one meter below the water's surface. This means the birds catch only surface fish or those that frequent shallow flats and shorelines – an important limitation, indeed a key fact of Osprey ecology. This limitation helps to explain Osprey nesting distribution, their preference for shallow bays and lakes, as well as their diet. Ospreys are hawks only slightly modified for fishing surface waters. Although great opportunists, they cannot compete with a bird like the streamlined booby, adept at plunging deep for fish, nor with loons or cormorants that swim underwater, pursuing their prey far into the murky depths.

5.1 The prey

With rare exceptions, Ospreys catch and eat live fish only. One bizarre exception was noted on Tiran Island in the northern Red Sea, where a few Ospreys learned to drop Conchs (*Lambis truncatus*) onto a large steel drum filled with concrete (Leshem, 1984). The birds had good aim and the shells broke readily, allowing the birds to eat the mollusks within. Equally unusual was a Florida Osprey I saw carrying a small Alligator (*Alligator mississipiensis*) back to its nest, a flashback to prehistoric times when *Archaeopteryx* snatched small lizards from the shores of Jurassic seas. Other people have recorded Ospreys with other nonfish prey: snakes, aquatic mammals, voles, squirrels, and even birds (Wiley & Lohrer, 1973; Proctor, 1977; Thorpe & Boddam, 1977; Taylor, 1986). Many such records are questionable, however. Ospreys regularly scavenge carcasses for nesting material, so they need not have killed all that they carry or all that one finds in their nests. Whatever the case, live fish have comprised over 99% of the diet of every Osprey population studied to date. Recently dead and dying fish are scavenged on occasion (Dunstan, 1974; Poole, 1984), but fish kills are rare enough that Ospreys must depend on live prey.

Identifying the species of fish Ospreys eat and determining their size are not so easy as one might think. Using binoculars or a telescope, an observer can often identify prey while a bird feeds, but accurately sizing a fish at a distance is much more difficult. One solution is to collect prey remains at nests or below feeding perches. As feeding Ospreys rip fish apart, they often drop the larger bones, especially the opercula (gill covers), fins, and tails, all of which can be collected, sorted, and compared with reference collections from fish of known sizes and species (Brown & Waterston, 1962; Swenson, 1978; Prevost, 1982).

Most studies of Osprey diet have used this method; unfortunately it has certain biases. For one thing, some fish are eaten more completely than others. Another problem is that scavengers may carry off remains before one gets to them. This was why Prevost (1982) installed wire baskets under Osprey feeding perches, returning periodically to collect the fish remains. This was an excellent precaution, but no other study has used baskets and few have verified prey remains by watching feedings. Thus our knowledge of Osprey diet is still imperfect.

Despite limitations, most studies of Osprey diet during the last two decades have agreed on a few key findings. First, Ospreys are opportunists, rarely choosy. If a fish is abundant, accessible, and the right size, it seldom goes unused. For any one population, however, only two or three species of fish usually meet these criteria. Along the south coast of New England, for example, about half the fish Ospreys eat during the breeding season are Winter Flounder (*Pseudopleuronectes americanus*), while herring (*Alosa* spp.) and Menhaden

(*Brevoortia tyrannus*) each supply another 20% of the diet (Poole, 1984). Herring and Menhaden, appearing seasonally in large schools at the surface of estuarine waters, are easy prey for Ospreys. Herring are also easy prey as they migrate into shallow streams on their way to freshwater spawning grounds.

Although these species are heavily utilized by Ospreys, they arrive in unpredictable pulses. Ospreys may be quick to cash in on the bonanza, but the schools inevitably move on, forcing Ospreys to fall back on Winter Flounder, their staple, until the next pulse arrives. Flounder are bottom fish, superbly concealed against sand and mud. Ospreys scoop them out of shallow waters, probably keying in on movement or perhaps the telltale puffs of mud that flounder leave behind when feeding (Prevost, 1977).

Inland Ospreys are likely to eat the same fish throughout the breeding season (Häkkinen, 1978; Swenson, 1978), but coastal populations change prey regularly, probably because most marine fish migrate seasonally. In eastern Nova Scotia, Ospreys catch Pollock (*Pollachius virens*) and herring when these fish move into coastal waters each spring, and flounder in summer when the others disappear (Fig. 5.1). These flounder are small and hard to catch (Table 5.1), so it is not surprising that Ospreys ignore them when possible. Comparable seasonal shifts were found in northeastern Nova Scotia where some Ospreys nest 8–12 kilometers inland and

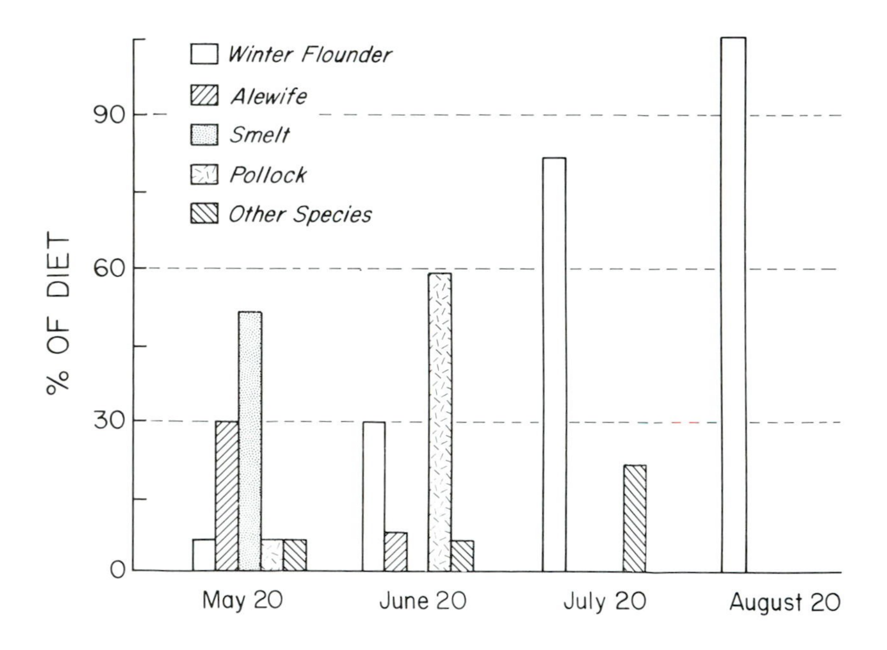

Figure 5.1. Seasonal change in Osprey diet along the east coast of Nova Scotia. Table 5.1 shows why some species figured heavily in the diet, even when others were available. Data from Greene *et al.* (1983).

Table 5.1. *The relative benefits of different fish to Ospreys in Nova Scotia (see Fig. 5.1). Data shown are mean values.*

Fish species	Weight (grams)	Time per catch (min.)	Grams caught per minute
Alewife	220	26	8.5
Smelt	54	22	2.5
Pollock	630	12	52.5
Winter Flounder	102	23	4.4

From Greene *et al.* (1983).

feed on nearby herring and suckers (*Catostomus* spp.) until midsummer when these fish move to deeper water (Prevost, 1977). Breeding males must then commute 20–25 kilometers round trip to hunt flounder in coastal estuaries, increasing their time away from nests by 80%–155% (Jamieson, Seymour & Bancroft, 1982).

Even within regions, habitat has a major influence on the size and species of fish Ospreys catch. At least 10 different fish species figure prominently in the diet of European Ospreys wintering along the coast of Senegal, although at any one location the birds take just a few species (Prevost, 1982). Mullet (*Mugil* spp. and *Liza* spp.) are a particularly important part of the Osprey diet there, as they are for most Ospreys feeding along tropical and subtropical coasts. Mullet swim in tight schools, frequent shallow water, and are rich in fats – an irresistible combination for a fish-eating bird.

On a broader geographic scale, Pike (*Esox lucius*), Bream (*Abramis brama*), Roach (*Rutilus rutilus*), Carp (*Cyprinus carpio*) and trout (*Salmo* spp.) all nourish Ospreys in Europe, with dietary differences depending on latitude and the size and depth of lakes fished (Cramp & Simmons, 1980). Likewise, Ospreys in western North America eat suckers, Carp, bullhead (*Ictalurus* spp.) and Perch (*Perca flavescens*) when nesting near warm, shallow lakes or reservoirs, but trout when nesting near deeper, colder waters (Swenson, 1978; VanDaele & VanDaele, 1982; Flook & Forbes, 1983)). In short, Ospreys are opportunists; they concentrate on fish that are most available. Whether they also select fish of particular species and size is a more difficult question to answer.

In the western United States, one study found that surface nets, used to sample at random, caught nearly the same species of fish that Ospreys did (Flook & Forbes, 1983). Another study found that net and bird catches differed substantially (VanDaele & VanDaele, 1982). In the latter case, however, a few species were especially vulnerable to Ospreys, either because they basked at the surface or were newly released from hatcheries. We need other such studies before we can know if Ospreys in general are selective about the species of fish they catch.

Regarding size preference, Ospreys in Senegal took slightly larger herring than nets did (Prevost, 1982), but Ospreys in other populations may not do the same. What is clear is that Ospreys concentrate on fish weighing 150–300 grams (0.3–0.7 pounds), although larger and smaller fish are said to be taken (50–1200 grams: Cramp & Simmons, 1980; Prevost, 1982). Measured another way, most fish that Ospreys eat range in length from 25–35 centimeters (about 10–14 inches), depending on region and season (Green, 1976; Swenson, 1978; Fig. 5.2). Fish weighing more than about 400 grams

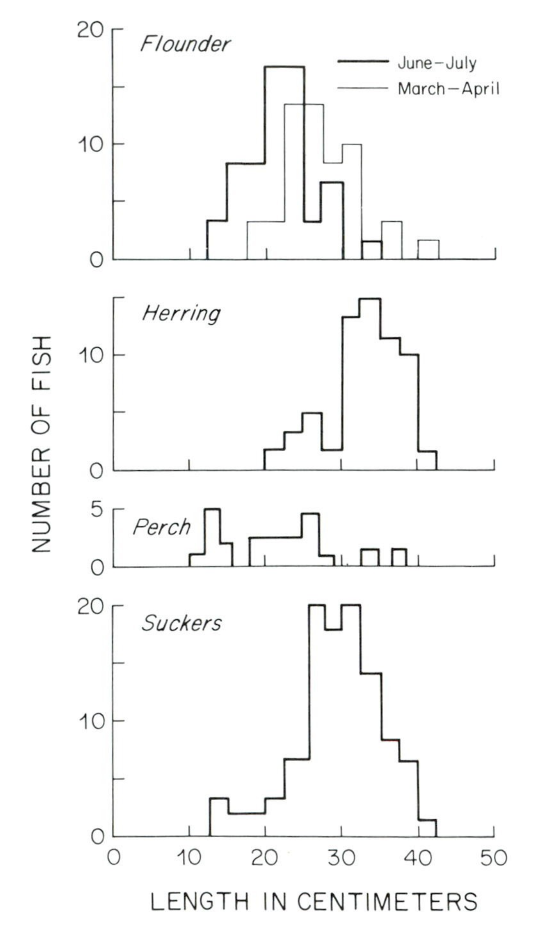

Figure 5.2. Length distribution of various fish caught by Ospreys breeding in southern New England (*top:* Poole, 1984) and in Nova Scotia (*bottom three:* Prevost, 1977). Fish of the same length but different species rarely weigh the same, so length can be a deceptive measurement if one is trying to determine how much food an Osprey consumes.

(25%–30% of a male Osprey's weight) may be difficult for the birds to carry away. Ospreys can release fish underwater (Rüppell, 1981), casting doubt on rumors that the birds sometimes drown because they lock onto fish too big to handle.

While large fish provide obvious benefits to an Osprey raising young, single birds can eat only about 300 grams at one feeding. Uneaten remains are dropped, or carried about until digestion makes room for further nibbling. In hot climates, fish spoil quickly so the birds may never finish a large carcass; one sometimes finds half–eaten carcasses in summer nests, swarming with insect larvae and exuding a potent stench. In cooler climates, an Osprey will sometimes clutch a carcass all day, feeding sporadically, reminding one of an old man unwilling to part with his cigar butt. Individuals sometimes settle in at dusk carrying fish that are subsequently consumed during the night.

5.2 Foraging adaptations

Superficially, Ospreys look much like hawks that hunt birds and rodents, but 10–15 million years of evolution have specialized Accipitriform morphology for catching, holding, and eating fish. Consider the Osprey's feet and legs, the bird's hunting weapons. Short, sharp spines cover the base of the foot pad and toes, an aid to gripping slippery prey (Fig. 5.3). The talons, long and razor sharp,

Figures 5.3. Close up view of Osprey foot and talons.

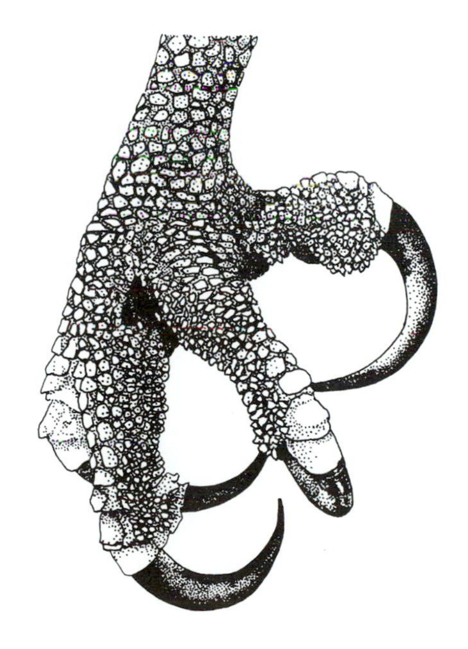

snap shut in 2 / 100 of a second (Rüppell, 1981), so fast that this might be a tactile reflex instead of a voluntary one. The flexible outer toe reverses position and lets the bird grip parrot-like with two toes forward and two back, providing extra dexterity and stability while the hawk subdues a fish's contortions. And long, unfeathered tarsi let an Osprey reach down into the water at the end of its plunge (Fig. 5.4).

The strong, hooked, compact beak is built for pulling, tearing, and twisting bites of fish loose as the carcass is held down against a perch. Watching an Osprey struggle to rip apart tough skinned, bony fish like flounder, one begins to realize that strength, leverage and a cutting edge are needed here, all of which the beak provides. Yet Ospreys will sometimes swallow entire tails, wads of skin, and formidable bones when confronted with a chunk they cannot rip apart. Their small intestine is long and narrow for a hawk, probably an adaptation to help digest tough scales and bones (Stone, Butkas & Reilly, 1978). Ospreys do regurgitate pellets – wads of indigestible scales and bones and sometimes grass (ingested incidentally) – but these are small and rarely seen, suggesting that most food passes through the gut.

Handling Ospreys, you quickly realize their plumage is dense and oily, which helps to keep them dry (Elowson, 1984). Yet repeated plunges or heavy rain will eventually soak these birds. They dry off

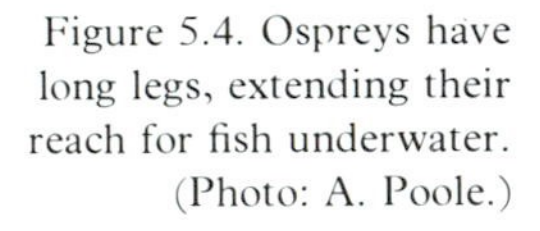

Figure 5.4. Ospreys have long legs, extending their reach for fish underwater. (Photo: A. Poole.)

by perching droop-winged and then slowly, methodically preening their feathers. In molting their feathers, Ospreys retain efficient flight, like most hawks but unlike many other birds. Ospreys molt their primaries gradually and in successive waves, starting at the innermost primary on each wing and working out to the tip (Prevost, 1983b). In this way, selection has ensured that only a few of each wing's aerodynamically important feathers are molted at once, and that adjacent wing feathers are never molted together. Molt stops just before migration and, for most male Ospreys, during the breeding season. Apparently hunting for a family demands the same peak flight efficiency that migration does.

5.3 The hunt

Most Ospreys hunt on the wing, actively searching out prey rather than quietly waiting at perches for fish to swim within striking distance. Flight burns about 10 times more energy than perching does (King, 1974), so the birds undoubtedly prefer to sit and wait for fish, although they can rarely do so efficiently. Even under ideal conditions – where lakes or streams have wooded shores, shallow margins, and a run of spawning fish – a perched bird can watch only a limited area. Thus it usually fares better on the wing, despite the extra cost. Wintering birds, as we saw in Chapter 4, tend to perch-hunt more than breeders do, perhaps because they have only themselves to feed and so can afford to wait longer for each fish caught.

Although it is difficult to tell when a perched Osprey is actually hunting, it is obvious with birds on the wing. Instead of rowing along in direct flight, fishing Ospreys fly slowly, sometimes circling back on themselves and often pulling up briefly to hover before moving on again, apparently stalking their prey. No one knows exactly what stimulates a hover. Hovering birds constantly scan the waters below, so spotting a fish or just seeing likely habitat could prompt this action.

The dive itself is spectacular, a quick release of tension built up during the preceding minutes. A diving Osprey tucks back its wings and abandons itself to the pull of gravity, usually falling steeply but maneuvering subtly with wing and tail all the while, to keep on track toward its target (Fig. 5.5). If unsure about a fish, the hawk may drop down in gradual steps for a closer look before making its final plunge or veering off to another area. Dives are sometimes aborted before hitting the water, the bird swooping up at the last moment if its prey proves elusive or undesirable (Fig. 5.6).

Try landing a fish 15%–30% of your own weight with your bare hands and you will begin to appreciate the problems of an Osprey lifting its struggling prey from the water. Even if the bird's talons strike deep, most fish are tough, die slowly, and struggle violently. To

Figure 5.5. The plunge: an Osprey diving for fish. The action shown here was drawn from a series of single frames from a movie film, part of a comprehensive analysis of Osprey hunting techniques using slow motion cinematography by Rüppell (1981). Note how the feet strike the water first with the head lined up directly behind, giving the bird an accurate sight on its target.

combat this, Ospreys often rest briefly on the water after diving, probably securing their prey, and then reach high with long, fast, almost horizontal wing strokes that start well above the tail and sweep down and forward of the head to eye level (Rüppell, 1981). The birds seem to gain much of their lift with the outer tips of their wings, taking off slowly like helicopters with heavy payloads. Once airborne, an Osprey usually rearranges its prey, one foot ahead of the other, so the fish's head points forward and its body is tucked close to the bird. This cuts wind resistance and speeds flight back to nest or perch. Most Ospreys fly low when bucking strong headwinds, hugging the surface of land or water and avoiding the worst blasts by doing so. If wet from diving, the hawks nearly always shake off in midair, twisting rapidly from head to tail and looking for all the world like a wet dog drying off after a swim.

Ospreys do most of their hunting five to 40 meters above the water, but this and other aspects of foraging behavior vary with the fish

Figure 5.6. Ospreys are not successful on every dive. This bird has just missed its target, a carved and painted wooden goldfish set out as a lure. This fine photo was shot in 1914 with relatively primitive camera equipment (Photo: H.H. Cleaves, Archives, Staten Island Institute of Arts and Sciences.)

pursued. When hunting fast–moving surface fish like mullet and herring, for example, the birds often fly close to the water, diving at low angles without hovering or fluttering down to snatch fish from the surface, barely wetting their legs (Prevost, 1982; Simmons, 1986). In Senegal, by contrast, Prevost watched Ospreys circle up high, 100–300 meters above the water, when searching for schools of sardines (*Sardinella* spp.) and flying fish (*Cheilopogon* spp.) several kilometers offshore. After spotting a school, the birds dropped down again before diving. Such offshore fishing is unusual for Ospreys, but it shows how alert they are to the sudden appearance of fish. Most often, hunting Ospreys hug coastlines or follow streams, venturing out only over small bays, lakes, and inlets.

Ospreys usually hunt alone, but small groups will form where food is especially plentiful. On Florida Bay, I have watched these birds, six or eight at a time, hovering together over schools of mullet that Porpoises (*Tursiops truncatus*) had driven into shallow water. Brown Pelicans (*Pelicanus occidentalis*) and various gulls often joined the diving melee, a pleasing kaleidoscope of mammal, bird, and fish that might have adorned a Greek vase centuries ago. Other Ospreys, such as those hunting over Nova Scotia estuaries, also flock at times (Prevost, 1977). These birds sometimes arrive at foraging sites together and lone birds often move towards diving birds, apparently expecting fish where others find them. Such behavior led Prevost to speculate that social foraging boosts an Osprey's hunting efficiency.

Nesting in colonies might also help Ospreys find fish. To test the idea that Osprey colonies are 'information centers' (Ward & Zahavi, 1973) – that breeders follow neighbors to foraging sites – John Hagan (1986) glued tiny radio transmitters to the central tail feathers of 16 male Ospreys at a colony in North Carolina and then tracked these birds with a receiver as they foraged. Eight of these breeding males quickly bit off their radio antennas (revenge?), but eight others ignored theirs and yielded some interesting data. Individual foragers, Hagan found, nearly always arrived back at nests from the same direction they had departed. Apparently these birds had just one place in mind when they left, and were not sampling different points on the compass (Fig. 5.7). Three sites, each about 14 kilometers from the nesting colony, turned out to be primary foraging grounds. Individual males stayed remarkably faithful to one site for the two months the study lasted, and two males tracked the following year were still commuting to the same foraging sites they had used the year before. In fact, about 75% of the colony's males departed on their hunting trips alone, usually circling up high (on calm days) and then making a beeline towards one site.

Where Ospreys forage close to a colony, individuals have a better chance to use the information their neighbors unwittingly dispense when returning with fish. Erick Greene (1987), who studied a small but dense colony of Ospreys nesting along the coast of Nova Scotia, found that individuals often departed on foraging trips soon after neighbors arrived back with fish. Not only did these departing birds

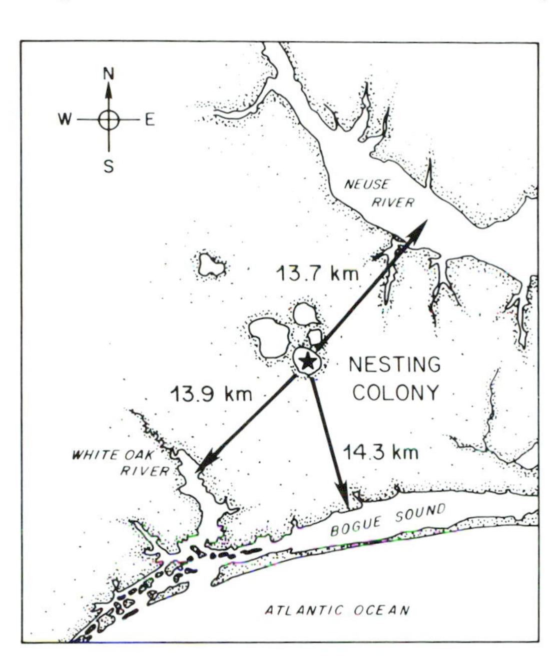

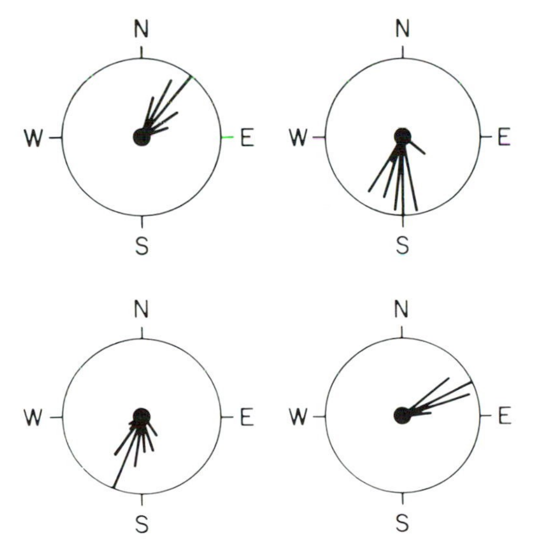

Figure 5.7. Ospreys breeding at Lake Ellis Simon (shown as star on map), North Carolina, foraged at three different coastal sites, each about 14 kilometers from their colony (arrows show direct routes flown by foraging birds). The four compass charts below the map show the directions flown by four radio-tagged males while on foraging trips from this colony. The direction most often chosen is shown as a radius of the circle, and all other lines are proportional in length to that one. Other males in this colony showed a similar preference for just one foraging site. From Hagan (1986).

tend to fly in the 'right' direction, the direction from which their successful neighbors had arrived, but most departed only when their neighbors arrived with certain species of fish. Schooling species like herring and Pollock elicited the greatest response. Nearby schools are generally easy to locate if an Osprey knows the right direction; the information is fresh. When neighbors caught solitary, randomly dispersed fish like flounder, on the other hand, Ospreys rarely bothered to leave the colony. One flounder caught is no guarantee that others will be in the same place. Thus, where fish are close by, Osprey colonies do sometimes function as 'information centers' and individuals apparently are able to sift out information that is useful to them. There is no indication, however, that such information has influenced Osprey breeding success.

Ospreys tend to forage most actively during the early to mid-morning hours and again in late afternoon, but other schedules are not uncommon. Where the birds must travel long distances to feed, three or four feeding pulses may occur regularly each day (Prevost, 1977 and 1982; Poole, 1984; Hagan, 1986). Given such regularity, tides usually (but not always) have less impact on feeding activity than time of day (Prevost, 1977; Ueoka & Koplin, 1973).

5.4 Hunting efficiency and energetics

To know how efficiently an Osprey hunts, one must be able to measure at least four parameters: the distance a bird travels to reach its foraging grounds, the availability of fish once it arrives there, and the size and species of fish caught. The first two measures help in estimating the energy expended in hunting, while the last two let one calculate energy gained – the number of calories released as a fish is consumed in the long, slow burn of an Osprey's digestion. To maintain itself and to raise young, of course, the energy an Osprey expends cannot exceed for long the energy it gains in metabolizing its food.

Two useful measures of foraging efficiency, therefore, are the energy (calories) gained per minute hunting and the amount of hunting time needed to meet the daily requirements of a bird or its family. These provide common coin for comparing different individuals and populations whose prey may vary greatly in availability and quality – boniness, edibility, or fat content, for example. Unfortunately, few studies have examined foraging Ospreys with energetics in mind, in part because these parameters, easy to measure with captive birds in the laboratory, are more elusive under field conditions. Instead, most studies have focused on how quickly and easily Ospreys catch fish: the percentage of dives that are successful and the amount of time needed per catch. We shall look first at Osprey hunting success, at why some hunts are quicker than others, and then go on to integrate success with energetics in an effort to determine how quickly and easily Ospreys meet their daily needs.

Osprey foraging success varies greatly among populations. Depending on locale, as few as 30% or as many as 90% of the birds' dives may produce a fish (Prevost, 1982). In an elegant comparative study, Jon Swenson (1979b) synthesized data on Osprey dive success from 14 regions of North America and found a strong correlation between the ecology of a particular species of fish and the Osprey's ability to catch it (Fig. 5.8). The hawks were most successful catching fish that ate slow moving benthic (bottom-dwelling) organisms like worms and clams and were least successful catching piscivorous (fish-eating) fish. Piscivorous fish, of course, must be quick and maneuverable to ambush their own prey, and this may partly explain why Ospreys find them more elusive than others. But habitat probably makes a difference too. Benthic fish, flounder for example, usually feed in shallow water where their only escape from a plunging Osprey is to scoot forward hugging the bottom. Trout and other piscivores, by contrast, generally frequent deeper water where escape lies

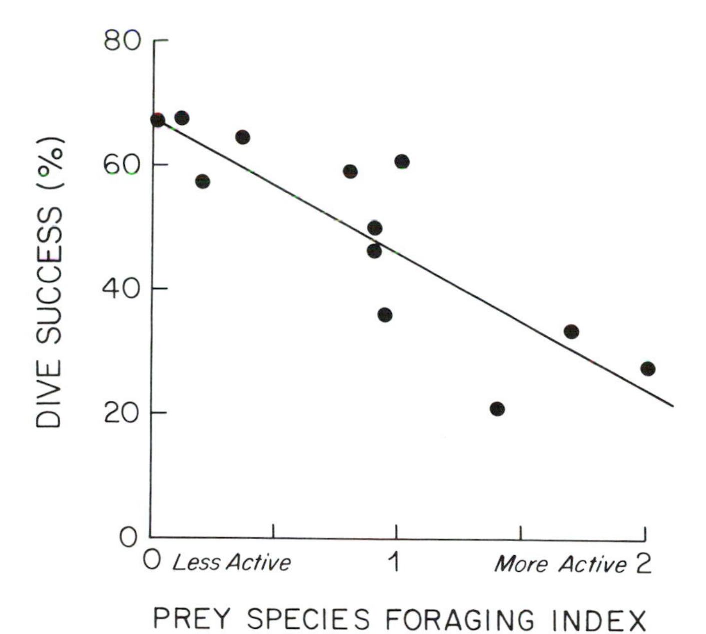

Figure 5.8. Average dive success for Ospreys in different regions of North America, in relation to the feeding ecology of their prey. Fish that eat mostly benthic organisms were given a foraging index of 0, those that eat plankton an index of 1, and those that eat other fish an index of 2. From Swenson (1979b).

downward as well as to the side. Small wonder that Ospreys thrive best in shallow water habitats.

Not surprisingly, older Ospreys hunt more successfully than younger ones; hitting a small moving target while diving is a tough skill to learn. Up to the age of six months, young Ospreys dive with only about half the success of adults, so their hunts usually last two to three times longer (Szaro, 1978; Prevost, 1982). Bad weather probably exaggerates these differences, although no one has checked this.

Weather does affect an Osprey's hunting performance, but studies have disagreed as to just how much. Tom Grubb (1977) watched Ospreys fishing at a lake in Florida and noted that cloud cover and waves both reduced the birds' catch. The Ospreys hovered and dove less often under such conditions. Grubb suggested that spotting fish below a darkened or rippled water surface was difficult for the birds. In Nova Scotia, however, where Ospreys hunted flounder over shallow-water flats, weather had little effect on dive success or the amount of hunting time needed per capture (Prevost, 1977). Likewise, studies at Osprey nests have shown that rates of food delivery rarely change with the weather (Green, 1976; Stinson, 1978). Thus weather apparently has more of an impact on Osprey foraging success in some habitats than in others. Such contradictions are probably due to regional differences in water depth, water turbidity, or the vulnerability of different fish.

Storms, however, are likely to reduce Osprey foraging success in all regions. Nestling Ospreys often starve during prolonged storms (section 7.3), so their parents must have difficulty finding food under such conditions. In South Africa, Ospreys simply stop hunting during high winds (Simmons, 1986). On the shallow flats of Florida Bay, storms churn up mud and turn the waters a turbid milky-green, forcing Ospreys inland to hunt over calmer, clearer ponds where fish are small and often hard to catch (Poole, 1984). These adjustments, and the others discussed above, are just a few of the most obvious ones that hunting Ospreys make as storms and gentler weather patterns swirl through their lives. Closer observations will no doubt show that changes in hunting locations and behavior are subtle and continuous.

Rarely will two different fish be equally valuable to a hunting Osprey. Fish differ not only in size but in the quality of their flesh and in the ratio of edible flesh to indigestible guts, bones, and fins. Twelve species of fish caught by Ospreys along the coast of Senegal varied from 0.7–1.2 kilocalories per gram of flesh and from an estimated 38%–65% inedible parts (Prevost, 1982). Soft-bodied fish are generally eaten more completely than those with harder skeletons, while differences in energetic value hinge largely on the lipid (fat) content of the flesh. Even within species, therefore, individual fish may vary greatly in nutritive value depending on age, sex and condition.

Spawning fish, packed with fat-rich eggs and sperm, are especially nutritious. Ospreys relish them. In New England, Ospreys abandon watery spawned-out flounder for the softer, richer, egg-burdened herring when these return to spawn each spring. It is amusing to watch an Osprey eating fish eggs, biting into the soggy gelatinous mass, smearing its beak with excess, and wiping this off onto a nearby perch. Like Melville's harpooners in *Moby Dick*, such birds 'dine like lords; they fill their bellies like Indian ships all day loading with spices.'

Knowing the size and species of fish Ospreys catch, the energy they derive from their prey, and the time spent hunting the fish, it is possible to calculate foraging efficiency and energy budgets for different populations (Appendix 3). Nonbreeding Ospreys wintering along the coast of Senegal, for example, usually catch one to three fish a day – an average of about six kilocalories per minute hunting – with efficiency at different locations ranging from three to ten kilocalories per minute hunting (Prevost, 1982). An Osprey expends roughly one kilocalorie for each minute hunting and only about one-tenth that for each minute resting, so most Senegambian Ospreys easily meet their energy requirements by hunting 20–30 minutes each day, about 3%–5% of the daylight hours – hardly a demanding schedule.

Consider, by contrast, a male Osprey breeding in New England that must feed its mate and three young as well as itself (Appendix 3). This bird's foraging efficiency is remarkably similar to that of his Senegal counterpart, but he expends a great deal more energy to keep up with family demands. As a result, he must hunt six to seven times longer each day than the wintering bird, twice as long just to supply the demands of his own body. Flight is an expensive activity, and 20%–25% of his daytime hours are spent on the wing.

How efficient does an Osprey need to be? Clearly, for a wintering bird feeding only itself, efficiency is not that critical. Even if Ospreys in Senegal had to double or triple their time spent fishing in order to meet daily requirements, they would still be loafing compared to most breeders. For breeding males, however, foraging efficiency is a real issue. These birds, spending nearly one–third of the day hovering, diving, and lugging fish back to their nests, are approaching their energetic limits, eating barely enough to balance the energy they expend hunting food for nestlings (Appendix 3). If forced to hunt longer, they would inevitably lose weight and condition, jeopardizing their chances of surviving a long migration. Energy budgets are not simply an academic exercise, therefore, but reflect true hunger and exhaustion, the means by which these birds are gauging limits to the different activities their lives demand of them.

In the next two chapters, we turn to Ospreys at the nest: their reproductive behavior, especially their choice of mate and nest site (Chapter 6), and their breeding rates, why some pairs raise more young than others (Chapter 7). In many ways, as we have seen, nesting is the most demanding and vital phase of life for an Osprey. Families require extra food, all of which must be found near the nest. Reproductive stakes are high. Pairs that succeed at breeding are those that are represented in the generations ahead.

6 Nest sites and breeding behavior

As we approach (the coast) by train or motor . . . we almost immediately begin to see the great nests of the Fish Hawk here and there in the tree tops and hear the peculiar whistling notes of the birds, so different from the cries of our inland birds of prey, and we realize that we are approaching . . . that delectable borderland where the elements of land and sea meet and intermingle.

W. Stone (1937)

I enjoy thinking of the aging naturalist Witmer Stone rolling through the pastoral landscapes of southern New Jersey in some 1930's roadster, dust billowing gently behind him, reveling in each new Osprey nest he passed. Yet his experience of seeing and hearing Ospreys at nests is not that unusual. Even with today's speed and noise, friends regularly tell me of spotting Osprey nests from the windows of passing trains and cars. They are conspicuous structures, usually large and high enough to stand out against the horizon from a distance. For the local populace, they become landmarks.

During the breeding season, Osprey nests are centers of intense activity, a stage where much of the drama of reproduction is played out. Because so much of an Osprey's reproductive effort is focused on finding and defending a nest, it is difficult to separate discussions of breeding behavior from those of nest sites. This chapter considers key aspects of Osprey breeding behavior, but especially those influenced by the size, location, and vulnerability of Osprey nests, and by the availability of sites to support them.

We probably know more about Ospreys at the nest than about any other aspect of their lives. Naturalists have spent countless hours staring at nesting pairs through telescopes and binoculars, peering into the details of their lives and consolidating the data that emerged. The study at Loch Garten (Scotland) is especially noteworthy. There, for over two decades, volunteers guarding Ospreys kept a detailed log of the nesting birds' activity (Green, 1976; Chapter 11). Cramp & Simmons (1980) relied on these notes in their scholarly treatment of the species; curious readers are referred there for more detail. My goal here is not to describe every aspect of Osprey breeding behavior but to concentrate on features most likely to be seen by someone who sits down to watch a nest for a few hours – an enlightening pastime. Ospreys live conspicuous yet unhurried lives. Observing them offers a respite from our own more frantic ones, a chance for self-reflection while we learn about another species.

6.1 Arrival: choosing a nest site

At northern latitudes in Europe and the United States, the arrival of Ospreys each spring corresponds with the break-up of ice and the movement of fish into shallow, sun-warmed waters. Like spring's first swallow, the first Osprey sighted is welcome proof that winter has loosened its grip. In New England, Ospreys return as herring move in from the sea and migrate up shallow rivers to spawn. In colder climates, at mountain or boreal lakes, Ospreys sometimes arrive and lay eggs before lakes have thawed, apparently finding fish in open water at nearby rapids (Henny, 1977b; Clum, 1986; Saurola, 1986). Late thaws, however, can delay laying for these populations as a whole (Wetmore & Gillespie, 1976), so probably only pairs with nests near rapids breed early. In subarctic regions like Finnish Lapland, the ice-free period may be too short in some years to allow Ospreys to breed successfully (Saurola, 1986); young cannot mature in time to migrate before fall freeze-ups.

This link with spring temperature accounts for latitudinal clines in Osprey arrival dates at north temperate latitudes (Fig. 6.1). Chesapeake Ospreys, for example, arrive and breed a week or two earlier than those in New England; both breed about six weeks earlier than Ospreys in Labrador. Temperature differences probably also explain why Ospreys breed earlier in Scotland, warmed a bit by oceanic waters, than they do in frigid Labrador at about the same latitude. The farther north one goes, the less predictable are spring thaws and thus Osprey arrival dates. In New England and Britain, the first Ospreys are nearly always sighted during the last week of March and the first week of April, respectively.

Males tend to arrive a few days ahead of females, but not predictably. Instead, older experienced breeders (male or female) generally arrive first, with younger breeders drifting in a few weeks later (Poole, 1985). Typically, a few days pass before these birds actually settle in at nests. If the weather is cold and windy, recent arrivals often seek shelter on the ground or behind windbreaks, perhaps not surprising for a bird that spends much of its life at tropical temperatures and has a relatively high limit of thermoneutrality (the temperature at which a bird must expend energy to keep warm) (Wasser, 1986). I recall a freak New England snowstorm in early April that drove Ospreys from their coastal nests into sheltered woodlots two or three kilometers inland, where they stayed, hunkered down, while the two-day storm blew over.

Established pairs nearly always return to their old nest sites, but new pairs, or those that lose a nest between seasons, must find new sites upon arrival. Such birds sometimes spend weeks trying to find an adequate site. It is not unusual to see sticks and clumps of grass dangling from the tops of trees in Osprey country, abandoned

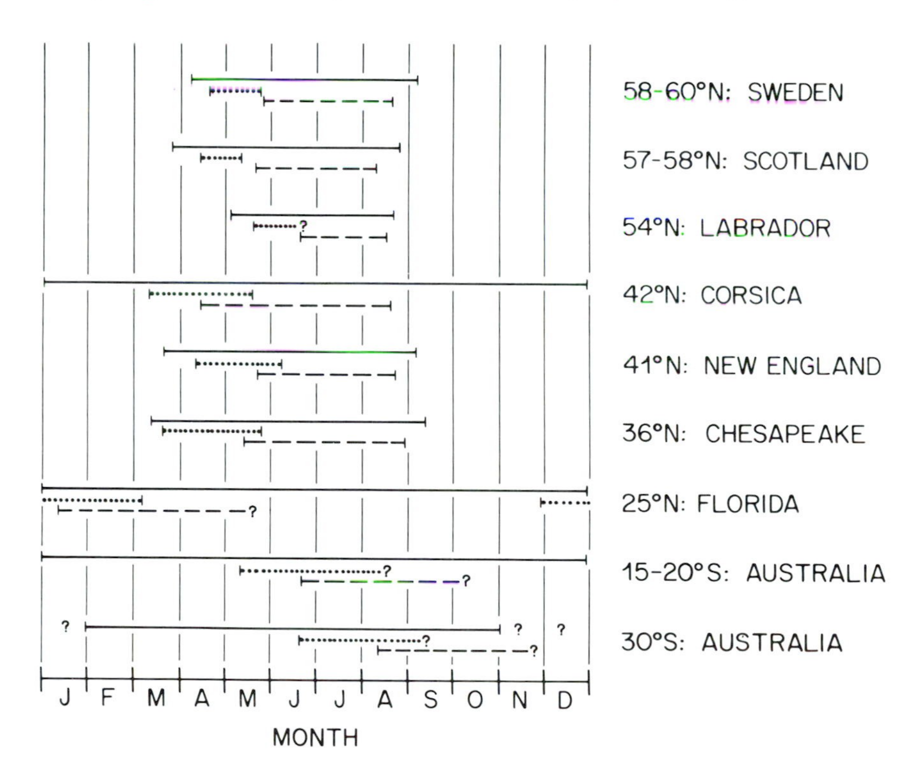

Figure 6.1. Breeding seasons for Ospreys in different populations. Solid lines show periods when adults are at nests. Dotted lines show laying periods. Dashed lines show when young are in the nest. Data are from the following sources: Sweden: Cramp & Simmons, 1980; Scotland: Green, 1976, and Dennis, 1984; Labrador: Wetmore & Gillespie, 1976; Corsica: Bouvet & Thibault, 1980; New England: Poole, 1984; Chesapeake Bay: Reese, 1977; Florida Bay: Ogden, 1977; Australia: P. & J. Olsen and G. Clancy, unpublished.

attempts at nest building. Few natural sites will hold a nest. Even in mature forests, less than one in a thousand trees may be suitable for the large stick nests of eagles and Ospreys (Newton, 1979).

What do nest-seeking Ospreys look for? What parts of a landscape match the search images stored in their brains? Since good nest sites are critical to breeding success and may last generations, selection undoubtedly favors individuals that discriminate carefully. First, Ospreys nearly always nest near food (near water, that means), although there are exceptions (Fig. 5.7). Three to five kilometers from water is often near enough. Second, the area around the nest must be open, giving the birds clear access when landing. The Osprey's long narrow wings are poorly adapted for maneuvering in tight quarters, so the tops of isolated trees, often dead ones, are preferred nesting spots (Fig. 6.2). If an Osprey does choose a live tree, it is generally flat topped, making nest building easy. Ponderosa pines (*Pinus ponderosa*) in the western United States, Tupelo (*Nyssa sylvatica*) in the eastern United States, and mangroves in the subtropics are all favorites for just this reason.

Figure 6.2. Osprey nests are usually placed in open sites, like the top of this dead tree. This allows the birds easy access to the nest and good visibility. In addition, height protects the birds from climbing predators. (Photo: M. Male.)

Artificial structures – power poles, radio and light towers, buoys, and special platforms – are fast replacing trees as preferred sites in many areas (Figs. 3.2, 9.1, 9.3, 10.1 and 10.3). They too provide a wide stable base for nests and they are increasingly plentiful. In addition, most artificial sites are difficult for predators to climb, a third and critical nest site requirement for Ospreys. This search for safety explains why so many Ospreys nest high in trees and on inaccessible rock pinnacles and cliffs. Curiously, Ospreys nesting in some regions, Scotland for example, have never adapted to cliffs. Perhaps there are local nesting traditions in different Osprey populations, although the range of sites used by any one population suggests that individuals are remarkably adaptable.

Water protects Ospreys from predators even better than height does. Overwater nest sites, in fact, have an almost magical attraction for the species. Bendire (in Bent, 1937) describes a memorable nest atop a rock spire in the midst of a huge waterfall. There 'the Ospreys reared their young amid the never ceasing roar of the falls directly below them.' Less dramatic overwater sites seem equally attractive. I have watched pairs spend days trying to build a nest on a single narrow offshore piling that could barely support five sticks. It is hardly surprising, therefore, that nesting Ospreys congregate where channel markers and offshore duck blinds (Fig. 6.3) abound, and in wooded swamps. Most climbing predators, like Raccoons (*Procyon*

Figure 6.3. Ospreys favor overwater nest sites because most ground predators are unwilling to swim to them. The nest seen here rests on the frame of a duck-hunting blind, a typical nesting site along the shores of Chesapeake Bay. (Photo: A. Poole.)

lotor), seem reluctant to swim far (or perhaps they are unaware of nests they cannot smell), so only aerial predators like owls reach overwater nests easily.

Islands, of course, provide overwater protection as well. Ospreys seem magnetized by islands, usually choosing smaller ones that ground predators have never colonized. Without the threat of predators, island pairs can often nest close together in low trees or even on the ground (Fig. 6.4). (It would be interesting to know if ground nesting is trial and error or if Ospreys sense safety before they settle in at ground sites.) Historically, the largest, densest colonies have formed on islands (Chapters 8 and 11). Even today, it is safe to say, at least half the Ospreys in the world are island nesters. In short, predators have exerted a major impact on the nest sites Ospreys choose, and thus on their breeding distribution as well.

Where possible, Ospreys like to nest near other Ospreys. This attraction may arise because some individuals try to usurp established nests instead of building their own or because the presence of established pairs hints that an area is suitable. In addition, good nest sites are often clustered – in swamps, on islands, or on power pylons, for example. In such situations, pairs often breed colonially with nests only 50–100 meters apart (Fig. 3.2), although loosely spaced colonies are more common.

Figure 6.4. On islands free of mammalian predators, Ospreys often nest on the ground and large breeding colonies can form. This nest was built on Gardiner's Island, NY (USA) where, at the time the photo was taken (1910), 250–300 pairs nested every summer, most clustered in just a few areas of this 12 km^2 island. (Photo: H.H. Cleaves, Archives, Staten Island Institute of Arts & Sciences.)

Such high density is possible because, as fish-eaters, Ospreys have no need to defend a feeding territory. Fish are a mobile and transitory resource, seldom found in predictable locations. Terrestrial habitats, by contrast, have a more stable prey base. Thus it pays most upland birds of prey to defend a feeding territory well beyond the nest site. But unlike birds that always breed colonially (gulls and terns, for example), Ospreys can and do nest successfully far from other Ospreys. Most Ospreys, in fact, are solitary nesters with nests often tens or hundreds of kilometers apart.

6.2 Pre-laying: courtship and nestbuilding

No natural sound is more characteristic of Osprey country than the rhythmic, penetrating *eeeet–eeeet–eeeet* of displaying males, a sign that the breeding season has begun. Squint aloft and you will likely find the source of that sound: a lone male, legs dangling, white head and belly gleaming in the sun, parading conspicuously overhead in a slow, high, undulating flight. This aerial display, called the 'fish-flight' or 'sky-dance', is often flown after successful hunts with a fish dangling conspicuously from the male's talons. Nesting material will suffice, however, and many males display while carrying nothing at all. Single males display most often, although mated males do so early in courtship.

The displays, which usually occur during fair weather and near the nest, seem to be of two intergrading types. One is the undulating flight, in which the male dives and rises repeatedly with stiff wingbeats in a shallow wave-like pattern, punctuating the top of each rise with a brief hover. The other is a pattern in which that brief hover is held while the bird rises and falls like a yo-yo (Cramp & Simmons, 1980). William Brewster, the early American ornithologist, described these aerial displays as 'the characteristic love-flight of the Osprey . . . not unlike that performed by several other species of hawks' (in Bent, 1937). In fact, courtship is probably only one function of this display. Predators, an intruding Osprey, or nearby humans will often provoke the hover display, so it seems there are territorial or threat overtones to this behavior as well. A single display can have different effects in different contexts, of course, and what constitutes a threat to an intruder could well be, in another instance, an advertisement to a lone female.

The Osprey's aerial displays, however conspicuous and dramatic, occur only sporadically. Feeding and mating are the main business of courtship. As in many other birds, female Ospreys are fed almost exclusively by their mates prior to laying (this is called courtship feeding), starting a long period of dependency that lasts until young fledge or the pair fails. After settling at nest sites, the females direct begging calls (Fig. 6.19) toward their mates or, if poorly fed or unpaired, toward any nearby male. If males fail to respond, females will hunt on their own, although they become increasingly reluctant to do so as egg laying approaches.

When a male does arrive with food, he generally finds a perch away from the nest, settling down to his portion of the fish (the head and anterior sections) before delivering the remainder to the female at the nest. Hungry females beg vociferously while males feed at nearby perches (rarely do females approach these males), but it is not clear what impact this and other begging has. I have watched males consume entire fish, head to tail, while their mates begged frantically nearby. Such behavior casts doubt on a current notion (Cade, 1982) that female raptors, larger than males, are dominant at the nest and thus able to force their mates to deliver food on demand.

During courtship, females in older pairs of Ospreys are fed more than those in younger pairs (Fig. 6.5). New males seem reluctant to transfer fish to their mates, especially as pairs form early in the season. This reluctance may take the form of 'mantling', whereby males face away from their mates and droop their wings over the food, seemingly protecting it. Well-fed females are less likely to beg from strangers or to copulate with them, suggesting that one key function of courtship feeding is to ensure mate fidelity (Poole, 1985).

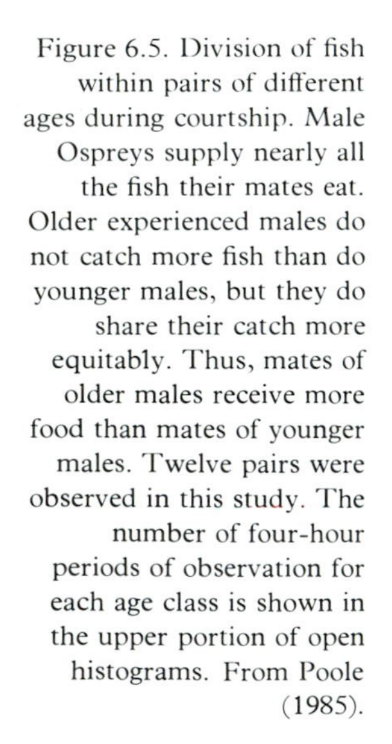

Figure 6.5. Division of fish within pairs of different ages during courtship. Male Ospreys supply nearly all the fish their mates eat. Older experienced males do not catch more fish than do younger males, but they do share their catch more equitably. Thus, mates of older males receive more food than mates of younger males. Twelve pairs were observed in this study. The number of four-hour periods of observation for each age class is shown in the upper portion of open histograms. From Poole (1985).

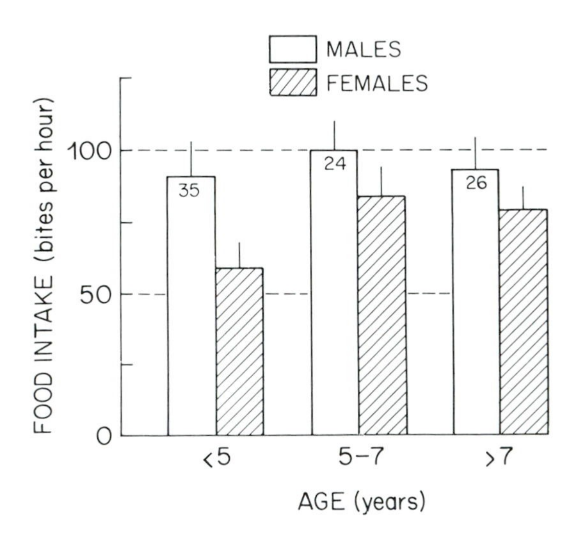

Table 6.1. *The average age of Ospreys paired with individuals in different age classes. Data from Westport, Massachusetts (USA)*

Age class (years)	No. of pairs sampled	Mean age of mates (years) ±1 standard error
3–6	39	5.6±0.3
7–12	45	9.5±0.5
>13	14	6.9±0.6

A. Poole, (unpublished).

In addition to feeding its mate, helping to keep her satisfied and close to the nest, a male often takes the extra precaution of guarding her during the few days before egg laying when she is most likely to be fertile. A guarding male follows his mate whenever she flies, sometimes swooping low over her back and touching her lightly with closed talons, as if urging a return to the nest.

Studies of courtship feeding in terns suggest that females may choose mates that feed them well (Nisbet, 1977). This seems unlikely for Ospreys, however. For one thing, Ospreys rarely divorce, so few arriving females get to move among sites and compare mates. In New England, I found that only five of 150 breeding attempts among banded pairs involved a divorced bird – one with a former mate still living. All divorces involved pairs that had failed the previous year. Switching mates may be more common where nests fail or are destroyed often (Judge, 1983), but good sites are sufficiently limited in most areas that males are probably reluctant to leave once they have built a stable nest. Females probably leave only if their mate provides no food.

It seems most likely, therefore, that pairs continue together year after year because both members of a pair have a strong attachment to the same site (Fernandez & Fernandez, 1977). Female Ospreys undoubtedly look for single males that possess a good nest or nest site, ignoring most other criteria. Such males are a rare commodity, given the low mortality rates in this species (Chapter 8), so females cannot afford to be too choosy. There is a significant tendency for younger Ospreys to pair with birds of about their own age (Table 6.1). Individuals older than about 13 years also tend to choose young mates, probably because available (single) birds are most often young ones.

Once Ospreys settle at a site and start courtship feeding, mating can occur at almost any time or place. Males mount females gently, talons closed and tarsi resting somewhat precariously along the female's back (Fig. 6.6, top); vigorous flapping helps the male maintain balance. If the female is receptive, she tips forward to allow the male's tail to scissor under hers, facilitating cloacal contact and the transfer of sperm (Fig. 6.6, bottom left). She usually maintains this position briefly after the male dismounts (Fig. 6.6, bottom right). Successful copulations, those in which cloacae touch, apparently depend on this forward tilt. Unreceptive females refuse copulations by keeping a horizontal posture or by tipping back on their tails, so males just stand there flapping or slide off. Failed copulations are common. Young pairs are less likely to copulate successfully than old pairs (37% *vs* 72%: Poole, 1985), largely because young males are reluctant to transfer food, leaving their mates unreceptive to breeding.

Most copulations take place at the nest because females spend most of their time there, but they can occur anywhere. No elaborate ritual or display precedes mating. Females sometimes appear to solicit copulations by tipping forward with raised tail and drooped wings (Saïller, 1977), but this is subtle behavior and hard to decipher. At other times, it seems that males mount with no signal from their mates, flying in from behind (like Yeats' Zeus taking Leda) or fluttering up onto the female from the nest edge. Only 24 of 93 attempts at copulation by male Ospreys occurred during feedings (Poole, 1985), so apparently food is not the immediate stimulus for mating. Pairs mate most frequently just prior to egg laying. Earlier copulations may actually have little to do with fertilization, serving instead to test a mate's receptiveness (strengthen the pair bond) and to synchronize the development of gonads. This latter aspect is especially important because females probably arrive back at nests each year with ovaries only partially developed, then going through a period of rapid ovarian development. In this phase, ovaries gain five to 15 times their initial weight before descending the oviduct (Lofts & Murton, 1973).

Figure 6.6. Ospreys mating. *Top:* male approaches female from the air, landing carefully on his tarsi, talons closed. *Bottom left:* female tips forward, tail up, to facilitate the copulation. Male flutters to maintain balance. *Bottom right:* post-copulation. (Photos: E. Saïller.)

Unlike some raptors that merely scrape out a small nest depression on a gravel cliff ledge, Ospreys usually build large, stable nests of sticks and other material. Some nests are huge, the result of thousands of additions carried in one by one (Fig. 6.7). During the peak of nest building, prior to incubation, pairs can make over a hundred trips a day for nest material (a small demand on their time,

however, as most material is gathered within sight of the nest) (Green, 1976; Levenson, 1979). Both male and female help to build the nest, although males tend to collect more (and larger) material than females. Females arrange most of the material once it reaches the nest. Old nests are repaired after damage during the nonbreeding season. Some large, stable sites are known to have been used by generations of Ospreys for dozens of years (Bent, 1937; Brown & Waterston, 1962). Although many Osprey nests seem larger than necessary for three small eggs weighing just 60–80 grams each, most nests are exposed, often to gale force winds, so their stability depends on their size and weight.

Ospreys vary their choice of nesting material during the prelaying period. Sticks are brought first, often dropped onto a site after being forcibly broken from trees or snatched from the ground. As the nest base grows, the birds switch to flatter and softer material for the lining that will support the eggs – an evident shift of 'search images'. Just before egg laying, a steady stream of lining material arrives: strands of kelp and mats of seaweed, bunches of grass, and bits of

Figure 6.7. Nest maintenance: an Osprey delivers a stick to its nest along the cliffs of Corsica, with the Mediterranean sparkling in the background. (Photo: J.-F. & M. Terrasse.)

cardboard. Unfortunately, today's Ospreys are also likely to line their nests with discarded plastic bags, a potential danger because young are sometimes smothered by the plastic. Tangles of old fishing line are also prized nesting material, though this can ensnare the young, constricting blood flow and amputating limbs.

There is a bit of the pack rat to every Osprey. Some of the New England nests I study are near dairy farms. I often find corn stalks, hunks of dried cow manure, empty fertilizer bags, and discarded rubber teat holders from milking machines woven into these nests. Anthropologists would enjoy looking through Osprey nests for the refuse of local human cultures. Dolls, sections of TV antennas, hula hoops, remnants of fish nets, old flannel shirts and rubber boots, styrofoam cups and buoys, a broken hoe, plastic hamburger cartons, and bicycle tires are just part of the list I have compiled during 10 years of checking Osprey nests. In nests of this same region during the nineteenth century C.S. Allen (1892) found:

> Barrel staves and hoops . . . laths, shingles, parts of oars, a small rudder . . . cork and cedar net floats . . . a hemp rope 20 feet in length . . . a toy sailboat with sail still attached . . . a feather duster . . . a blacking brush and boot jack . . . part of an oilskin "sou'wester" . . . one rag doll . . . a small worn out door mat . . . bleached cattle and sheep bones, especially sheep skulls.

Most Ospreys build only one nest, although pairs that lose their eggs or young often build a second nest nearby. These alternative, or 'frustration', nests are sometimes used when the builders return the following year. Where nest sites are plentiful and failure rates high, 40%–50% of the pairs may possess (and guard) alternative nests (Judge, 1983), although many such nests are not viable. In New England, only 10%–20% of the population builds second nests. Such nests are important insurance. Pairs with second nests do not waste weeks building a new nest if their old one blows down during the nonbreeding season.

6.3 Polygyny

Most Ospreys are monogamous. Rarely, however, they breed as polygynous trios – one male breeding concurrently with two females. (In only one instance, where male mortality may have been abnormally high, was a significant portion of an Osprey population found to be polygynous (Fernandez & Fernandez, 1977)). From 1980–1986, I found three such trios out of 190 monogamous pairs that bred in the colony I study, a typically low incidence for the species. In each of the three cases (Table 6.2), the male involved had already paired with one female when another, unmated female settled at an empty nest nearby. Males all defended these second sites, usually against other males, and copulated regularly with both mates. Secondary females were short–changed, however; males rarely fed them and spent most time at their primary nests (Table 6.2). Not surprisingly, secondary females usually had to hunt for themselves and rarely hatched eggs.

Why would a species that nearly always breeds monogamously occasionally breed polygynously? And why would secondary females tolerate mates that rarely fed them? Lack of undefended, available nests could be one reason. All Osprey trios studied so far bred where nests were close together, at least within sight of each other (Fernandez & Fernandez, 1977; Mullen, 1985). Thus males were able to defend two sites almost as easily as one. In other birds, polygyny is thought to arise when a male defends sufficient resources (food or nest sites) to support two or more mates (Verner & Willson, 1966; Mock, 1983). Male Ospreys that defend two empty nests, each an attraction for a single female, are thus the most likely candidates for polygyny. Few males ever find an undefended nest close to their own, however, and this accounts for the absence of trios in most populations. Secondary females put up with polygyny, I believe, because their choices are limited. Those that leave may have trouble finding another empty nest. Those that stay may breed successfully if the primary female fails or dies.

In a slightly different version of Osprey polygyny, two females may very rarely share a nest with one male (Fig. 6.8). Bent (1937) recorded one clutch of five eggs and one brood of seven young, and Dennis (1983) reported two females trying (unsuccessfully) to incubate at the same nest. Most female Ospreys, however, are quick to fight others that come near their nests. Why would a few be tolerant? In other species of birds, similar sorts of cooperative breeding have evolved where resources were particularly scarce, causing family members to linger for years in the home territory before dispersing (Emlen, 1984). It may be that female Ospreys share nests only with sisters or daughters, and only where lack of food or nest sites is a real constraint.

6.4 Laying and incubation

In the Northern Hemisphere, most migratory Ospreys lay their eggs 10–30 days after arriving in breeding territory (Fig. 6.1). In any one

Table 6.2. *Time and effort invested in primary and secondary nests by three male Ospreys that were polygynous, and reproductive success at those nests. Each male bred and guarded two mates, and each mate had her own nest. Primary and secondary nests were close together. Secondary mates arrived and laid eggs later than did primary mates. Defense noted here was against other Ospreys. Each trio was observed for 30 to 40 hours during the courtship or incubation periods.*

	Nest	
	Primary	Secondary
% of day at nest	64	18
Defense (% of all attempts)	59	41
Nest material (% of all deliveries)	59	41
No. of fish delivered/hour	0.38	0.03
No. of young fledged/nest	1.67	0.00

A. Poole, (unpublished).

Figure 6.8. Very rarely, two female Ospreys will share a nest with a single male, as seen here. It is not clear why this happens, although shortages of food or nests may be factors. (Photo: Y. Eshbol, Israeli Raptor Information Center (IRIC).)

population, pairs that lay late are generally those that arrive late, although younger pairs often have longer courtship periods than older, experienced pairs and lay later (Fig. 6.9; Poole, 1985). Watching young pairs getting started, it is easy to see why they are slow to lay. Most lack nests to start with, and young males tend to be reluctant to feed their mates. Thus, young females probably lack two key stimuli for converting follicles into eggs.

While most migratory Ospreys lay eggs in spring, most resident populations lay in winter (Fig. 6.1). In south Florida, for example, Ospreys lay from early December until late February. Australasian Ospreys, including those at temperate latitudes, breed from May to August, during the austral winter. Although migratory Ospreys probably arrive and lay as early as spring temperatures allow, it is not clear why residents produce eggs during the short, cool days of the

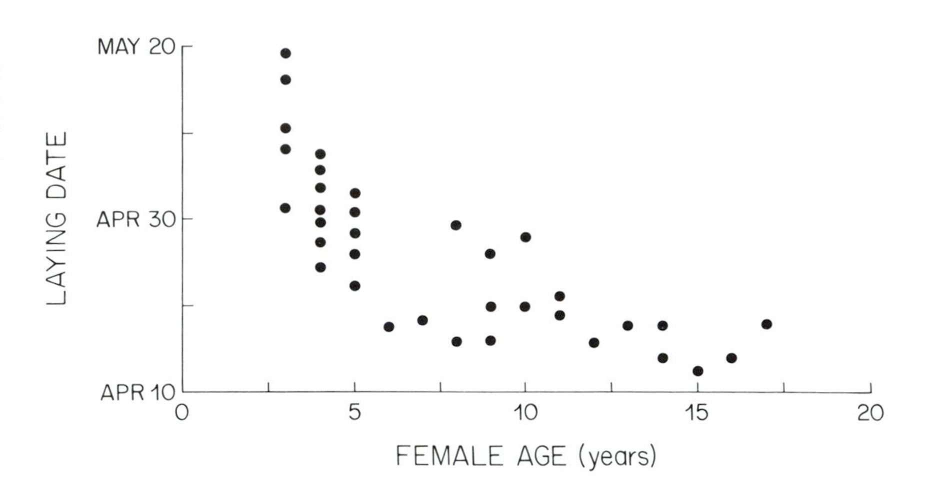

Figure 6.9. Laying dates for female Ospreys of different ages in southern New England. From Poole (1984 & unpublished).

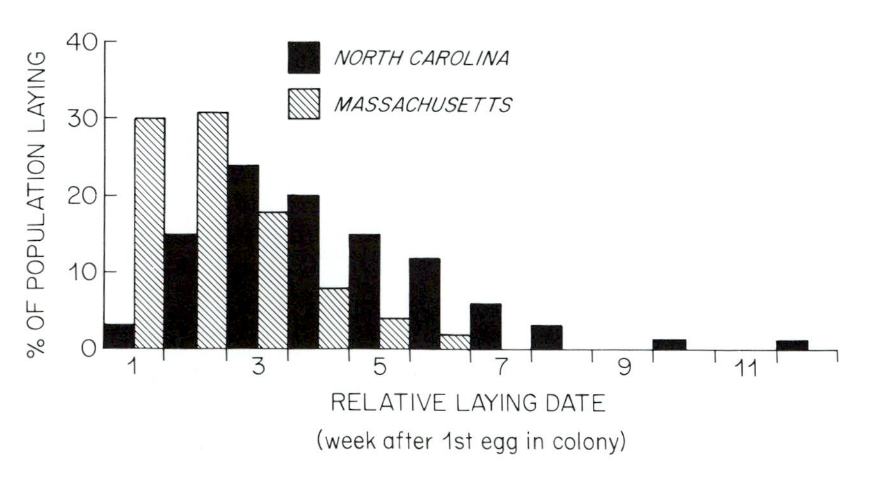

Figure 6.10. The distribution of laying dates in two different Osprey colonies. A total of 102 nests were sampled in Massachusetts over a four-year period (Poole, 1984) and 136 nests in North Carolina over a two-year period (Hagan, 1986).

year. Summer heat may drive fish from shallow waters, making them less available to Ospreys. Within populations, laying is most synchronous at northern latitudes and at inland locations (Figs. 6.1 and 6.10), probably because these Ospreys arrive close together and because late breeders, rarely successful, are heavily selected against. In most populations, early layers predominate (Fig. 6.10).

The eggs of the Osprey, wrote the noted American ornithologist and zoologist A.C. Bent (1937),

> are the handsomest of all the hawks' eggs; . . . a selected series of them is a great addition to an egg collector's cabinet. I shall never forget my envious enthusiasm when a rival boy collector showed me the first fish hawk's eggs I had ever seen. Nor could I ever forget the peculiar pungent odor that clings to these eggs after many years in the cabinet, a fragrant reminder of many hard climbs.

Bent describes these eggs in detail, emphasizing the variation in background colors (white to creamy white, shading into fawn, dusty tan, or vinaceous-cinnamon) and blotches (bay, chestnut, burnt sienna, and various shades of brownish drab and deep reds with grey undermarkings) that mark most eggs (Fig. 9.5). Little wonder that so many nests have been robbed to decorate the felt-lined drawers of egg collectors (Chapter 11).

Osprey eggs are about the same size and shape as those of large domestic hens. They are laid one to two days apart and, in any one clutch, later eggs are significantly smaller, on average, than earlier ones (Table 6.3). Other birds have a similar tendency to put slightly less energy into late eggs (Lack, 1954). Osprey eggs usually hatch in the sequence they are laid. Chicks hatched third or fourth, therefore, not only hatch from smaller eggs than their siblings, but more importantly, they are several days younger as well (Table 6.3). This age gap puts younger siblings at a significant disadvantage. At hatching, for example, third chicks are 25%–30% lighter on average than their older nestmates, and thus much less likely to survive, especially when food is scarce (Table 6.3; see below).

Watching a nest from a distance, it is easy to tell when incubation starts. Instead of standing at the nest edge, as pairs generally do during courtship, incubating birds quickly settle into the nest cup after returning, rocking slightly to bring the eggs into full contact with their brood patches – bare, heavily vascularized sections of skin on the breast. The nest cup is usually so deep that only the head and sometimes the back of the incubating bird are visible (Fig. 6.11).

Table 6.3. *Average differences within Osprey clutches in egg sizes, hatching asynchrony, and survival of eggs and chicks. Survival data are from two different regions: one where food was plentiful and one where food was scarce. Eighty three-egg clutches and 20 four-egg clutches were studied.*

Eggs: laying sequence	% laid that hatched much food – little food	% size of 1st egg
1st	83 – 89	—
2nd	83 – 77	98
3rd	78 – 42	94
4th	69 – 40	92
Chicks: hatching sequence	**% hatched that fledged much food – little food**	**No. days younger than chick hatched 1st**
1st	96 – 87	—
2nd	94 – 88	1.5
3rd	83 – 38	3.4
4th	27 – 0	7.0

A. Poole, (1982a, 1984).

Figure 6.11. A female Osprey incubating eggs. (Photo: A. Poole.)

Both males and females incubate, and both have brood patches, although females generally incubate longer and always take the night shift. Individual males show great variation in the extent of their participation, incubating about 20%–35% of the time at nests in Scotland (Green, 1976), 26%–57% in Montana (Grover, 1984), and 13%–66% in New England (Poole, 1984). It is not clear why such variation occurs. Males that incubate less do not deliver more food to their mates, nor do they spend more time hunting. They just perch more than other males do.

Usually a male delivers a fish before taking over incubation duties, but the switch can occur without a food transfer. Shifts can last only a few minutes, almost as if the bird relieved left unwillingly, although the mean duration of daytime incubation bouts at 10 nests I studied ranged from 29–75 minutes. Many bouts lasted four or five hours.

Most Osprey eggs hatch five to six weeks after being laid. The average incubation period for 33 Massachusetts pairs was 39 days, with a range of 35–43 days, very near the range of other Ospreys studied (Green, 1976). Whether such variation is due to efficiency of incubation, to inherent differences among eggs, or to other causes is not known. If their eggs are destroyed or collected early in incubation, a pair will usually lay a second, smaller clutch about three weeks later.

6.5 Nestlings

Ospreys enter the world weak, wet, and relatively helpless (Fig. 6.12, top). The strongest part of a newly hatched Osprey is its neck, which helps drive its egg tooth, the hardened tip of the upper mandible (beak), into the confining eggshell. If you handle a hatching egg, you are likely to hear the chick's faint peeps within and see the tapping of a persistent bill through a hole in the shell.

Most Ospreys hatch within one or two days of pipping their shells. Like other birds of prey, they emerge as 'semi-precocial' young. This means that down covers most of their body, that their eyes open hours after hatching, and that they can actively take food from their parent's bill. In fact, the best way to tell if Osprey young have hatched is to watch for feedings. A female parent with chicks remains at the nest after the male delivers a fish, dipping her head low to rip off small bits of the prey and delicately presenting them to her chicks. Nestlings are still far from mobile at this stage and often hard to see from a distance, but they do not fear humans. They stand weakly, heads wobbling, begging at any movement near the nest edge and then collapse soon afterward in a huddle, seeking the warmth of surrounding eggs and nestmates.

Once dry after hatching, Osprey chicks are kept warm with their first down, a short, thick, buff colored plumage that is replaced 10–12 days later. As they age, the crop develops quickly, becoming a large

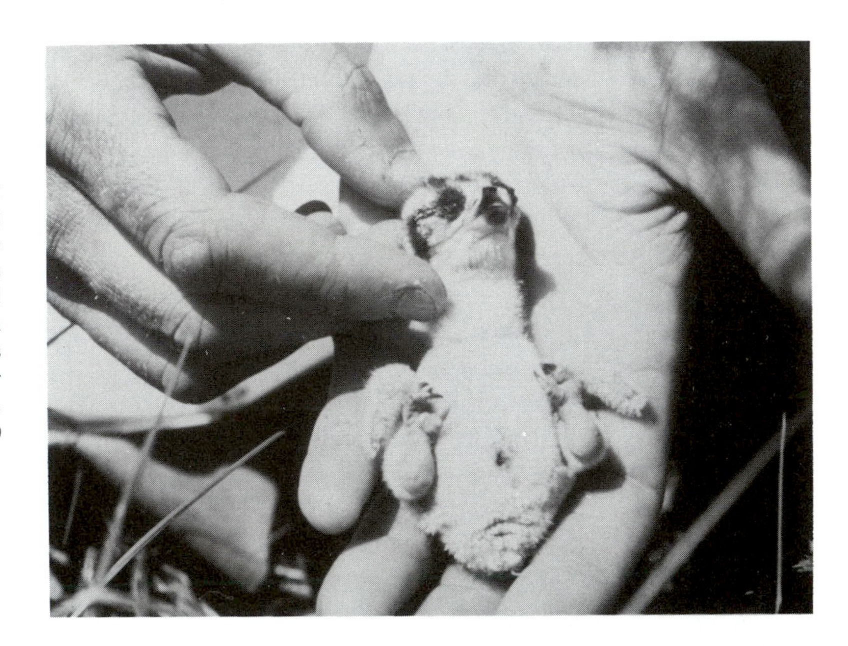

Figure 6.12. *Top:* newly hatched Osprey, about one day old. Note the white egg tooth on the upper beak and the scar from the umbilical (yolk sac) attachment along the belly. *Bottom:* young about 8–10 days old. Note how this bird's crop has expanded to allow for food storage. Without such an adaptation, Osprey chicks would need to be fed far more often than they are. (Photos: A. Poole.)

and conspicuous part of their body when full of food (Fig. 6.12, bottom). With the emergence of their dense, wooly second down, Osprey chicks enter their 'reptilian stage', which lasts another 10–15 days. They are black, scaly, and often crouch at danger, reminiscent of their reptilian ancestors. A conspicuous light tan streak runs down their spine. Their feet are bluish-grey and their claws long and black, colors that will hold for the bird's life. At 10 days of age, nestlings are already mobile, quickly approaching the female when food is delivered, fighting siblings when food is scarce, and backing up to eject their feces over the rim of the nest. Feathers start replacing down when the young are about two weeks old. First to appear are rusty-golden pinfeathers on the head and neck. Darker body feathers follow slightly later, and primaries, secondaries and rectrices (outer feathers of the wing and tail) emerge at about 20–25 days.

By the time Ospreys are 30 days old, they have already achieved 70%–80% of their total body weight and growth is slowing noticeably (Fig. 6.13). Developmental pathways, therefore, channel energy into Osprey bone and flesh ahead of feathers. Few studies have measured growth of nestling Ospreys, so it is hard to make reliable comparisons among different pairs and habitats.

Three findings are worth noting, however. First, the length of a young chick's culmen (bill), which is easily measured in the field, correlates well with age in growing young (Poole, 1982a). This allows one to pinpoint the age of a chick even when its hatching date is not known. Second, when chicks are 20–35 days old, females, which are heavier than males as adults (Fig. 2.2), add weight faster than males do (Fig. 6.13, top). Thus, body weight is a reliable criterion for sexing nestlings older than 30–35 days (Fig. 6.13, bottom). Interestingly, male Ospreys, having less weight to gain, may actually mature faster than females. In one study, young presumed to be females took longer to start hunting on their own than presumed males (Schaadt & Rymon, 1982). Third, young in regions with a poor food supply grow more slowly, on average, than young at well-fed nests and often fledge later (Poole, 1982a). Ospreys from nonmigratory Mexican populations, for example, fledged, on average, when 63 days old (10 broods; range = 52–76 days old; Judge, 1983), considerably older than Ospreys fledging in migratory populations (range = 50–55 days old) (Stotts & Henny, 1975; Stinson, 1977; Poole, 1984). Without the pressures of migration, resident populations may have evolved slower growth rates than migratory populations. Alternatively, Mexican Ospreys may just be nourished less well.

Osprey chicks that are persistently hungry often fight with siblings for access to their mother distributing food. Well-fed young almost never fight (Poole, 1979, 1982a). If undernourished, the dominant chick, usually the oldest and largest in the brood, reacts instantly when food is brought to its nest. It rears up, bill gaping, and lunges toward a subordinate chick, viciously pecking its head and back (Fig. 6.14). The victim may try briefly to defend itself, but more often it turns away, crouching submissively at the nest edge. The loser then waits until the dominant chick has fed. Once full, dominant chicks often allow smaller siblings to move toward the parent. During continual food shortages, however, dominant chicks are hungry so much of the time that subordinates rarely eat and slowly starve to death or are pushed over the nest edge. Parents never interfere in such squabbles, apparently because it fits their interests to raise one

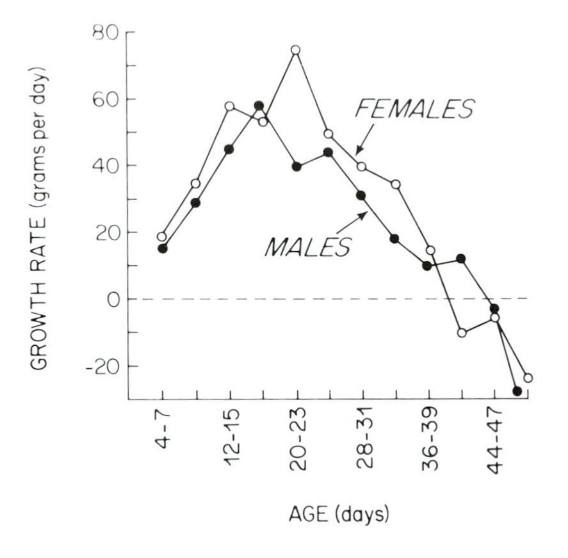

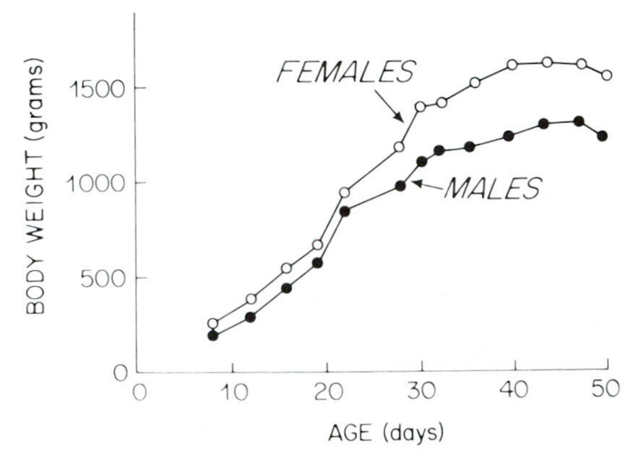

Figure 6.13. Growth of nestling Ospreys, five males and five females. *Top:* nestling growth rate. Note that female growth rates peak after that of males; females, that is, put on more weight, for a longer period, than males. *Bottom:* weight change during the nestling period. Note that most body growth has occurred by the time chicks are 30 days old, about two-thirds of the way to fledging. Data from McLean (1986).

or two well-fed young, instead of the three or four weak ones that would result if scarce food were shared equally (Lack, 1954; Stinson, 1979). Seeing runt or fighting chicks in an Osprey nest is thus a good indication that food is in short supply.

It takes fish to fuel the growth of young Ospreys, cold-blooded flesh ripped apart piece by piece to nourish warm blood, resilient feathers, keen eyes and light, hollow bones. Despite many studies of nesting Ospreys, there are remarkably few data showing how much fish these birds consume at different nests and in different regions. I have found it almost impossible to compare food delivery and consumption among different studies. To some extent, thousands of hours of observation at Osprey nests have been wasted because people lacked a common, accurate method of determining an Osprey's feeding rate.

Simply counting the number of fish males deliver is not adequate. Ospreys catch fish that range in size from 50–500 grams (Chapter 5). The main problem, therefore, is how at a distance to measure the size of a fish clutched in an Osprey's talons (Fig. 6.15). Some studies have done this by estimating the length of fish brought to nests and then converting that estimate to weight via specific length–weight

Figure 6.14. Hungry Osprey chicks fight for dominance and access to food. Aggression begins when chicks are about 10–12 days old.

regressions. Prey length, however, usually estimated through a telescope, is very difficult to judge accurately. In addition, small errors in estimating length mean large errors in weight because a fish's weight varies as the cube of its length.

One solution is to count the number of bites of fish each bird eats (Poole, 1985). This is easy to do because Ospreys are conspicuous feeders, dipping their heads low to reach prey held in their talons and pulling up to rip off each bite. Bites are then consumed by the parent or passed along to a chick (Fig. 6.16). The beauty of bite–counts is that one can quantify how fish are shared at the nest, and how this division changes during the breeding season (Fig. 6.17, top). Yves Prevost (1982) has suggested that studies attempting to estimate Osprey food consumption also should consider the caloric content of different fish because this varies widely among species. Ideally, therefore, future studies should report food consumption at Osprey nests both as weight (grams of fish) and as energy (kilocalories).

Measuring Osprey food consumption in bites, I found that the most fish was brought to nests during the nestling period and the least during incubation (Fig. 6.17, top). The most noticeable shifts in food consumption occurred as nestlings aged and, to a lesser extent, between courtship and incubation. Nestlings obviously need more food as they grow, although deliveries level off when young are about

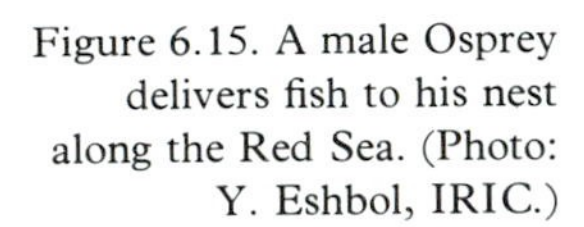

Figure 6.15. A male Osprey delivers fish to his nest along the Red Sea. (Photo: Y. Eshbol, IRIC.)

30 days old (Fig. 6.17, bottom). There is a curious drop in the number of fish delivered to nests just before young fledge (age 40–55 days) and a rise about 15 days after that (Fig. 6.17, bottom). Chicks also lose weight just before fledging (first flight) (Fig. 6.13, bottom); maybe their parents withhold food to encourage nest departure.

Depending on the size of her brood, the age of her young, and on her mate's hunting proficiency, a female generally receives only about 15%–20% of the food her mate catches (Fig. 6.17; top). Females thus seem willing to sacrifice their own food intake in order to feed their chicks. The male's share, however, varies much less. He always takes a larger percentage of the daily catch than his mate does. As the main provider, expending much energy hunting (section 5.4), the male's reluctance to sacrifice is not surprising.

During the last 10–15 days before fledging, young Ospreys regularly exercise their wings, developing the muscles that will power their first tentative flights (Fig. 6.18). This urge to exercise seems catching; nests with two or more young are often a pandemonium of beating wings. When not feeding, the female parent

Figure 6.16. Female Ospreys distribute fish that males bring to the nest, biting off small pieces of flesh, eating some themselves, and giving the rest to their young. Counting bites lets one determine how food is shared within the family. (Photo: Y. Eshbol, IRIC.)

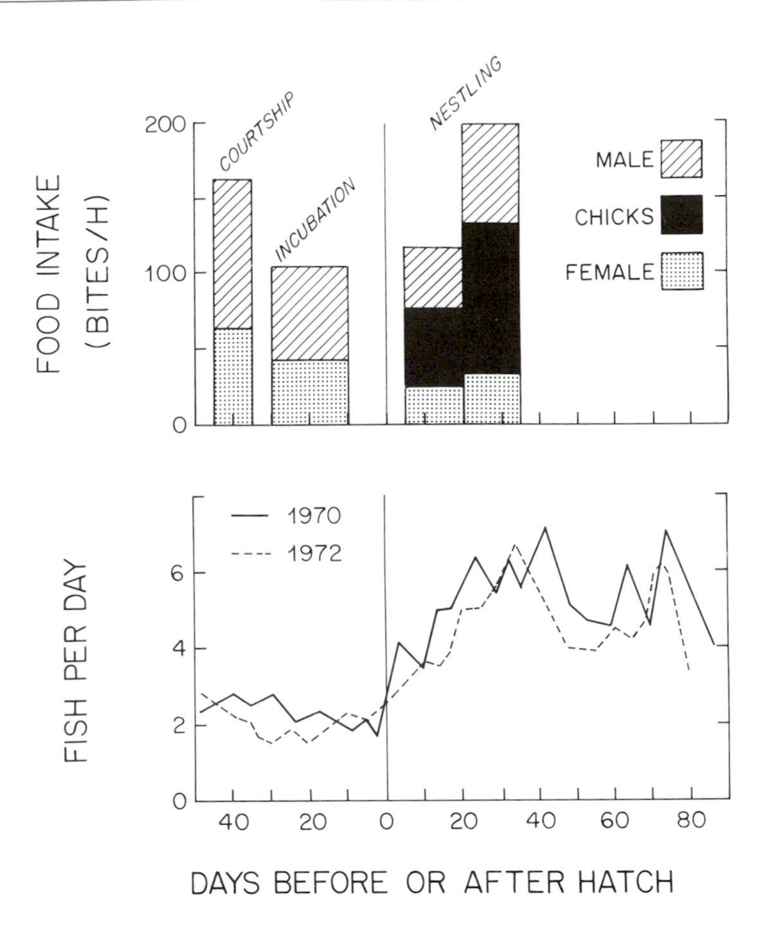

Figure 6.17. Food delivery and consumption at Osprey nests. *Top:* the average amount of fish consumed at different points in the breeding season by six Osprey families with three young each. Categories (male, female, chicks) are not cumulative. No data were available for more than 35 days after the hatch. By putting pre-weighed fish in nests, I found that bites weighed about 0.6 grams each, on average. From Poole (1984). *Bottom:* the number of fish delivered per day to the well studied nest at Loch Garten, Scotland. From Green (1976).

Figure 6.18. Young Ospreys become very active a week or two before fledging, regularly exercising their wings in preparation for flight. Those seen here are about 45–50 days old and will fledge in 5–10 days. (Photo: M. Male.)

often leaves the nest at this stage to perch nearby and avoid the confusion. As their proficiency develops, nestlings face the wind and jump repeatedly, wings pumping, legs dangling, and wild–eyed. Eventually, a puff of wind will catch one, dropping it over the lip of the nest and forcing its first awkward flight.

6.6 Post-fledging

The weeks between an Osprey's fledging and its independence (no food from the parents) are a critical period, one that should ease a young bird into the strength and competence it needs to complete dispersal or its first migration. In migratory populations, fledged young continue to depend on their parents for food at least 10–20 days (Stinson, 1977; Poole, 1984), although some catch their own fish only 2–3 days after leaving the nest (Schaadt & Rymon, 1982). In resident subtropical populations, where seasonal pressures are less intense, fledglings often stay near the nest and take food from their parents for eight or ten weeks (T. Edwards, personal communication). Within populations, one might expect families that fledge early to stay together the longest.

Fledglings are generally poor foragers (Szaro, 1978; Prevost, 1982), so food from their parents provides a vital backup. Fledged young beg persistently. When parents deliver a fish to the nest, one fledgling usually grabs and defends it, biting off chunks until satiated or until a sibling takes over the carcass.

I was surprised to discover how readily fledgling Ospreys leave their own nests to visit others nearby, sometimes switching nests many times a day before returning home (making the rounds, as it were) (Poole, 1982b). Parents seldom chase and often feed intruders, probably unable to recognize a foreign fledgling once it reaches their nest. Subordinate chicks are most likely to switch, usually seeking nests with younger chicks they can dominate at feedings. Such nest switching seems confined to colonies where nests are close together.

Successful independence hinges on a young Osprey's ability to catch fish. Many people have wondered how the young develop this skill, whether it is innate or taught by the parents. Colonel Meinertzhagen (1954), the imposing British soldier and naturalist, watched fledgling Ospreys while he vacationed at a Swedish lake and suggested that parents lure their young from the nest by flying past with fish. He also saw parents repeatedly drop fish into the water, apparently encouraging their young to stoop for them. Others have questioned these interpretations. Hoagy Schaadt and Larry Rymon (1982), for example, released hand-raised young to the wild and found that these birds hunted successfully three days to three weeks later, even though they had no parents. Finding such innate fishing behavior in Ospreys, of course, does not exclude the possibility that parents also encourage their young to start fishing.

Once young Ospreys can fly from danger, their mother can leave them to hunt on her own, helping her mate to provide for the brood. Not all mothers do so, however. While females delivered about 60% of the fish eaten by fledged young in Virginia (Stinson, 1978), studies of other populations have found that they rarely hunt for their fledglings (Green, 1976; Mullen, 1985). In many other species of raptors, females readily leave the nest to forage once their young are old enough to survive without brooding (Green, 1976; Newton, 1979). Why do Ospreys differ? Part of the answer may lie in the Osprey's open nesting habits. Even large chicks left unguarded could fall victim to daytime predators like eagles. Disturbance by other Ospreys could also be a problem. Nests left unguarded are a lure to floaters – unattached Ospreys looking for nest sites – and young could be attacked during attempted takeovers.

Once they are independent and feeding themselves, no one knows exactly where young Ospreys go. In resident populations, young gradually drift away from their nests, sometimes lingering weeks or months at nearby foraging sites. In migratory populations, some young may stay at nests a week or two after their parents depart, but most leave when their parents do (Stinson, 1977; Fig. 4.6). In New England, I often see them perched inconspicuously along the wooded edges of tidal rivers and inland ponds, taking their ease during the last few dog days of August. By early September, most Osprey nests are empty at northern latitudes, breeding activity over until spring warmth rekindles it. Only the nests remain, wistful reminders of a bird that pursues an endless summer.

6.7 Vocalizations and nest defense

Vocalizations are not just a way to communicate with mates and neighbors, they are also the first line of an Osprey's defense, an effort to warn intruders away from an occupied site. Breeders are most vocal, but others, even migrants, call when stimulated. My ears discern three types of calls: guard calls, alarm calls, and begging (Fig.

6.19). All have been described in detail by Yves Prevost (in Cramp & Simmons, 1980) and all are variable.

Breeders give guard calls when an intruder, usually another Osprey, approaches their nest. This is a slow call, a series of whistled notes falling rapidly in pitch, sometimes followed by two or three more intensive wavering notes when the intruder nears the nest. Prevost suggests that the whistled, down-slurred notes sound like a boiling kettle being taken off heat, 'tiooop-tiooop-tiooop'. Their slowness hints at laziness and lack of intensity, perhaps just a distant warning, a call of occupancy. Watching Osprey nests on hot summer

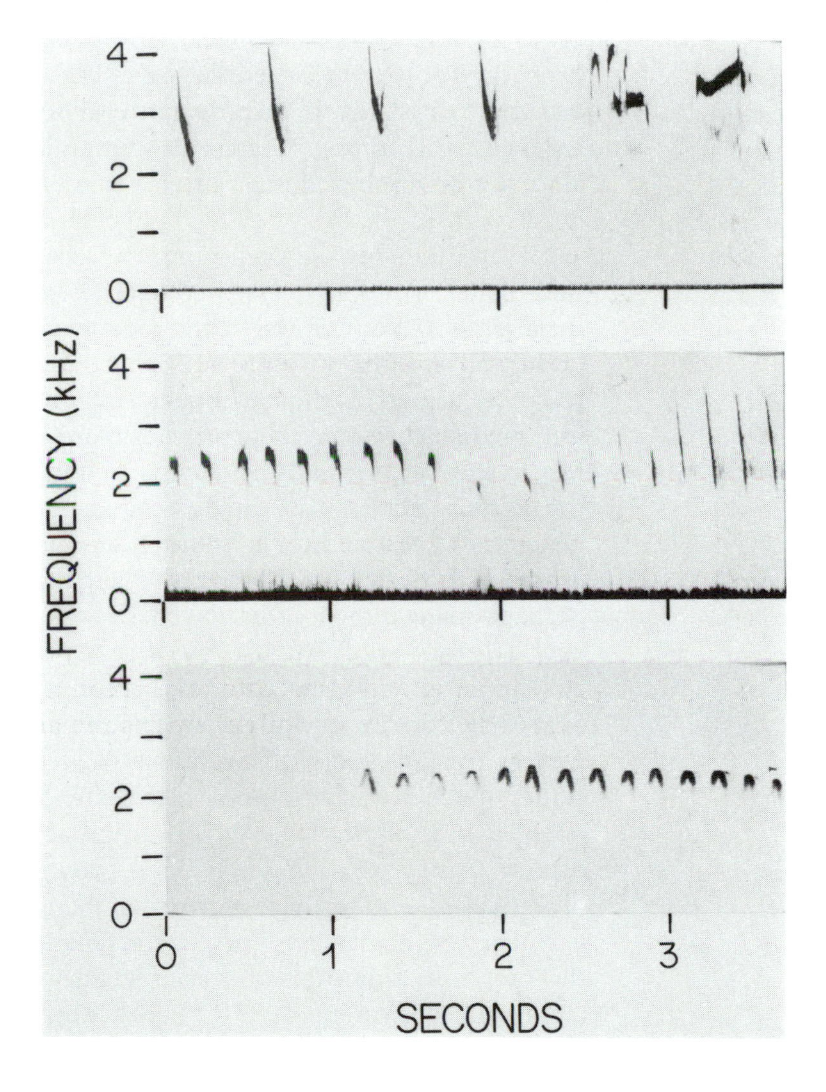

Figure 6.19. Sonograms showing brief portions of Osprey vocalizations. *Top:* guard calls, usually given by a perched bird when intruding Ospreys approach. *Middle:* alarm calls, usually given when a predator or a human approaches. Note the differences between the alarm calls of females (left) and males (right). *Bottom:* begging by a female Osprey. Only females and young beg. (A. Poole & T. Highsmith, unpublished.)

days, guard calls can almost lull you to sleep. These and other calls probably vary among individuals, allowing Ospreys to recognize each other by voice, as individuals of other bird species can do (Beer, 1980).

Alarm calls vary from short, clear whistles that fall in pitch to faster, more excited, high-pitched squeals given as threats approach. These squeals are the same calls given by males during their aerial courtship displays, so 'alarm' calls probably have multiple functions: attracting mates, claiming territory, and alerting neighbors and mates to danger (one bird's alarms often trigger those of others nearby). Male and female alarms differ slightly, as do their guard calls. Females have stronger, lower pitched calls than males, probably due to their larger body size (Fig. 6.19, middle). Alarm calls can grade into a harsh, rasping, gutteral 'ick-ick-ick' when the birds are highly threatened. It is these threats, along with the rapid squeal alarms, that are heard most often when one approaches an Osprey nest.

Nesting females, especially hungry ones, direct begging calls to their mates. In food-stressed colonies, begging calls often continue throughout the day. Eight or ten females sometimes beg at once, an orchestra of hungry Ospreys. The female's begging calls closely resemble those of the nestlings, suggesting that she is a vocal intermediary between them and her mate. Male Ospreys sometimes respond to begging calls by leaving to hunt or by delivering food to the nest, but just as often they are unresponsive. It is not clear why or when males respond, or if some individuals respond quicker than others. A thorough study of the give-and-take of begging in Ospreys, perhaps using taped calls played back to males, would make an excellent field study.

When a potential predator (an eagle or a human, for example) approaches an Osprey at the nest, pairs are likely to respond both with alarm calls and with conspicuous, stereotyped posturing: erect stance, back feathers raised, neck extended, and wings partially opened, beating slowly. A close approach by an aerial predator usually brings on a chase. Nearby ground intruders elicit power dives. Males tend to chase more often than females, especially at the periphery of the nest territory, while females tend to guard closer to the nest, but this division varies greatly among individual pairs (Jamieson & Seymour, 1983). There are certain males and females in the colony I study, for example, whose nests I avoid checking whenever I can. These are formidable nest defenders, very likely to strike with talons in passing dives. A blow to the head by a diving Osprey is a memorable experience, one not willingly sought by the sane; I very nearly lost an eye to such a bird.

Not all animals are intimidated by diving Ospreys, however. Pat Mullen (1985) described Canada Geese that usurped Osprey nests by arriving and breeding earlier than the hawks did. Ospreys spent days bombarding the nesting geese, but were never successful at regaining their old nests.

Intrusions by other Ospreys can be just as much of a problem for nesting Ospreys as the rare gull, eagle, or fox that passes by. Even at the isolated Loch Garten nest in Scotland, intruders – nonbreeders searching for nests – appear every two or three days, on average, often trying to land at the nest (Green, 1976). Intruders can be very persistent, and have been known to smash eggs while fighting for a nest (Harvie-Brown & Macpherson, 1904). Injury to breeders is a real possibility. One dead male I found near a nest in New England had been seen fighting another Osprey the day before. I suspect he was killed defending his nest. Even unpaired birds defend nests against other Ospreys of the same sex, an indication that fights are really over nests, not mates.

This chapter has provided a glimpse of Ospreys at the nest: how they choose a nest site, build and defend a nest, attract a mate, time their breeding, and raise their young. Knowing this, we can appreciate the difficulties a pair faces in transforming eggs into fledglings. The following chapter considers these difficulties in detail, focusing especially on variation in Osprey breeding success – why some pairs raise more young than others.

7 Breeding rates

The act of leaving offspring to succeeding generations is the most important aspect of any animal's life.
Perrins & Birkhead (1983)

Most individual birds are cautious parents.
Drent & Daan (1980)

Breeding rates are the speed at which Ospreys reproduce themselves, their ability to contribute offspring to the next generation. Breeding rates are the standard way to measure success in the lives of wild animals because reproductive success is the major form of competition confronting individuals. If you could read an Osprey's genes like a set of blueprints, you would be perusing designs handed down by steady winners in the game of reproduction, designs that had survived eons of winnowing in the strong winds of natural selection. The concern in this chapter is who those winners are, or why some Ospreys produce more young than others. To avoid complication, the focus here is on natural rates of reproduction, rates set by selection and unaffected (as far as we can tell) by pollutants or persecution. Those two influences are examined separately in Chapter 9.

Consider the hurdles an Osprey must clear to breed successfully: migrating to and from nesting grounds, finding a mate and nest site, producing and incubating eggs, and nourishing young to fledging and independence. One finds many facets to Osprey breeding, and thus many points where rates can be set. First and most obviously, a bird must cross a breeding threshold. Individuals may forgo breeding if they are late, undernourished, or unable to find a place to nest. Other controls on breeding rates include the number of eggs laid (clutch size), the number of eggs that hatch and survive as chicks (brood size), the number of fledglings that survive to breed, the age at which a bird starts breeding, and the number of years a breeder lives. All of these aspects contribute to lifetime reproductive success, some more than others. This chapter focuses on clutch and especially brood size because these have been well documented. Other factors controlling Osprey breeding rates have remained more elusive and can be considered only briefly.

'All birds of prey,' wrote Aristotle in the *Historia Animalium*, 'are unprolific, except the Kestrel . . . which drinks, and its moisture along with its heat are favorable to generative products.' Although Aristotle may have jumped to conclusions about Kestrels (probably *Falco tinnunculus*), his general statement shows characteristic insight. Like most other large predators, Ospreys reproduce relatively slowly. While some ducks and tits may raise a dozen young a year, Ospreys rarely raise more than three, many eagles only one. What holds these predators in check? Rather than the 'heat' and 'moisture' Aristotle thought aided Kestrels, we must look to food availability as one pervasive influence on Osprey breeding rates. While lack of nest sites often prevents breeding and limits population size, and weak or vulnerable nests often fail, it is hunger – young dying of starvation or nests abandoned by hungry parents – that accounts for most reproductive loss among Ospreys. Even when other causes of breeding failure are invoked, age or weather for example, hunger is usually the root cause.

Do Ospreys face such food constraints often? Do some individuals find it easier to bypass these constraints than others? How does an Osprey measure loss of condition – decide if and when to abandon a nest, for example? These are important questions because knowing the limits that individual birds face reveals how selection sets overall breeding rates for a species. Before tackling these questions, however, we consider how researchers define breeding rates for Ospreys and how they gather these data.

7.1 Methods and terminology

Osprey breeding success is usually reported in one of three ways: as the average number of young fledged per successful nest (brood size), as young fledged per active nest, or as young fledged per occupied nest. By definition, 'successful' nests are those that fledge young, 'active' nests are those where eggs are laid, and 'occupied' nests are those where pairs are present, regardless of whether or not they lay. Each method gives a useful perspective on breeding rates, but most Osprey studies report only active and successful nests because nonbreeding pairs are so elusive. Without frequent and careful checks at nest sites, one cannot tell if pairs assumed to be nonbreeders actually laid eggs but failed early, or if those seen at one site were counted earlier at another site nearby. In addition, breeding rates based on occupied nests alone provide no clue to success or failure. A low figure could be due to many nonbreeders or to the failure of pairs that did lay, an important distinction. Ideally, therefore, studies of Osprey breeding success should report the number of nonbreeders, layers, and successful pairs, as well as the number of young fledged at each nest.

Such accurate measures of breeding rates require at least four visits to nests. The first notes occupancy of nest sites and evidence of egg laying (an incubating bird is a good clue), while the second counts eggs (clutch size). Clutch size is best measured about eight to ten days after a bird starts incubating, enough time for a full clutch but before egg loss becomes a factor. A third visit should be timed to coincide with the hatch. Because first and last eggs may hatch six to eight days apart, and because young often die and disappear soon after hatching, extra visits during the hatching period improve the accuracy of counts. A fourth visit checks for young a week or two before fledging. By then, most runt chicks have died so one can predict with reasonable accuracy how many young a nest will fledge. Large young are easily visible from a distance, especially while they feed; patience and a telescope are all that are needed to count them. Aerial surveys facilitate such counts where nests are scattered, but it is difficult to see eggs and small young from the air.

Within Osprey populations, some pairs may start laying weeks or months after their neighbors (Figs. 6.1 and 6.10). As a result, visits timed to check the peak layers in an area often miss data on early and late pairs. To monitor an entire population, therefore, frequent visits to breeding areas are really needed. I monitor nests weekly in the New England colony I study, but such frequency is obviously impractical where nests are scattered or inaccessible. Most breeding surveys, in fact, have relied on just two or three visits to Osprey nests, some on just a single visit near the peak of fledging (Postupalsky, 1974).

Another problem confronting the Osprey researcher is how to identify individual birds. Leg bands (rings) are perhaps the best and simplest way to mark Ospreys. Every nestling Osprey banded receives a permanent aluminum leg band with a unique number engraved on it. Some nestlings are also given colored leg bands in unique combinations, visible at a distance with a telescope. But to identify an Osprey without color bands, to read its leg band number, one must trap it.

Trapping Ospreys sounds hazardous and tricky but it is often quite easy. The nest is the lure. The trap is a 'noose carpet', a wire dome whose surface bristles with short monofilament slip nooses. Placed over eggs or young in the nest, the trap is effective when a parent returns and walks on the dome, tangling its feet in the nooses. The trap is weighted so an Osprey can fly with it, but not too far. Once on the ground or in water, trapped birds may struggle occasionally but are easily approached (Fig. 7.1). The whole operation – trapping, banding, and release – goes quickly (30–45 minutes) in most cases, so disturbance is minimal. While one handles a trapped bird, its mate usually returns to cover the brood; the trapped bird itself is often back at the nest minutes after release.

There are other methods of trapping Ospreys, at their feeding perches for example (Prevost, 1982), but noose carpets are most effective for breeders. Trapped Ospreys can be color-banded for later identification by telescope. Most Ospreys are too smart to trap twice.

7.2 The effect of food supply on breeding success

Convincing evidence that food availability has influenced Osprey breeding rates is not so easy to come by as one might think. Just finding small broods or emaciated chicks in an area is no guarantee that Ospreys are short of food there. Disease or predation could be complicating factors. So far, most evidence of food limitation has been circumstantial, based on differences in breeding success among nests, areas, or years. The best such work has also measured the number and size of fish delivered to nests, showing that successful nests receive more fish than unsuccessful ones or that broods are small because chicks have died of starvation.

Regional differences in food supply

Although nesting Ospreys are supported by habitats as different as tropical lagoons, boreal forest lakes, and temperate estuaries, most pairs lay the same number of eggs. Three has been the most common clutch size recorded for the species, worldwide (Appendix 4). Differing percentages of two and four-egg clutches account for what

Figure 7.1. *Top:* this breeding Osprey was trapped at its nest using a 'noose carpet', seen trailing behind the bird. The traps have floats attached so a bird does not drown if it lands in water. *Bottom:* People trap Ospreys to band them or to identify individuals via numbered leg bands put on before the birds can fly. (Photos: M. Male.)

little variation occurs among populations. One-egg clutches are very rare and, when recorded, are probably often the result of counts made too early or late in the nesting season. Clutches with more than four eggs, even rarer than singles, are probably the efforts of more than one female (section 6.3).

Clutch size often changes with latitude and habitat in birds, although it is not always clear how this relates to food supply. Lack (1968) noted larger clutches at higher latitudes in a variety of birds, but exceptions are common (Perrins & Birkhead, 1983). Osprey clutch size likewise follows a latitudinal trend, but not uniformly. In North America, southern resident populations lay about 0.5 fewer eggs, on average, than northern migrants breeding along the mid-Atlantic and New England coasts (Appendix 4; Judge, 1983).

Small clutches may be adaptive because the surplus food needed to raise three or four young is seldom available at the subtropical latitudes where most residents breed. Rates of food delivery are low and brood reduction widespread in many subtropical populations, compared with most northern ones (Table 7.1). Although this appears counterintuitive – food should be plentiful where the climate is warm year-round – the very lack of seasonability is thought to account for the scarcity (Ricklefs, 1980). A surge of spring and summer growth provides surplus food for most temperate birds, whose breeding numbers are often reduced by migratory or winter mortality. (Spring fish migrations, for example, are a key source of food for northern Ospreys.) Tropical and subtropical populations, by contrast, generally live closer to the carrying capacity of their habitats because seasonal variations in mortality and food supplies are less pronounced there.

Within temperate latitudes, however, Osprey clutch size does not vary consistently with latitude. Populations breeding in Sweden, Scotland, and northern Michigan (USA), for example, lay fewer eggs on average than migrants at US coastal sites 10°–20° farther south (Appendix 4). It is tempting to attribute such differences to food availability. Most northern populations feed at boreal lakes whose waters are generally less productive than nutrient-rich coastal and estuarine waters (Valiela, 1984). Yet brood size at fledging, a good measure of an ecosystem's ability to support Ospreys, is not consistently smaller at inland sites compared with coastal sites (Appendix 4).

Compared with clutch size, brood size, which reflects survival of eggs and chicks, seems more directly linked to regional differences in

Table 7.1 *Osprey clutch size, brood size (young fledged per successful nest), and rates of food delivery to nests with young in different regions. Brood/clutch shows survival of eggs and young. Mean values are shown.*

	Colony			
	Florida	Baja	NY1	NY2
Latitude (°N)	25	29	41	41
Brood size	1.1	1.0	1.4	2.0
Clutch size	2.7	2.8	3.2	3.2
Brood / clutch	0.41	0.36	0.43	0.63
Food delivery (g / d)	816	550	890	1320

Data from Florida and New York (NY) are from Poole (1982a & 1984). Those from Baja California (Mexico) are from Judge (1983).

Figure 7.2. The male Osprey foraging for this brood apparently could not find enough fish to nourish all three of its young. Runted chicks like the one in the middle result when hungry, dominant nestmates prevent smaller siblings from feeding. Runts nearly always die of starvation. (Photo: A. Poole).

food supply. Rates of food delivery over a two-year period, for example, were substantially different in two US Osprey colonies (NY1 and NY2: Table 7.1) just 20 km apart, and brood sizes paralleled these differences. Checking nests in NY1 where food was scarce, I commonly found chicks fighting and younger siblings runted and half-starved (Fig. 7.2). Such sights were rare in well-fed NY2.

Further evidence that Osprey breeding success is linked to food availability comes from regions where these two values have varied together over time. In the once large and extensive nesting colony on Gardiner's Island (off eastern Long Island, NY), for example, Ospreys fledged large broods (two or three young, on average) during the 1930s and 1940s, when a nearby commercial fishery was booming; but small broods (0.7–1.5 young, on average) were fledged during the late 1970s, after that fishery had all but eliminated the colony's main food source (Spitzer, 1978; Poole, 1982a).

A comparable case involves Ospreys breeding along Chesapeake Bay, where water quality and fish availability have declined in recent years (Officer *et al.*, 1984). During the mid-1980s, broods on some Chesapeake creeks were far smaller than broods checked a decade earlier in the same locales, and many young were found dead or starving in nests (McLean, 1986; P. Spitzer, unpublished). Male Ospreys apparently had difficulty feeding their young. During 1985, males in one part of the bay spent more time away from nests, presumably much of this foraging, and delivered 35% less food to their nests, compared with males observed in this same area during 1975 (McLean, 1986).

Differences among pairs

Within local populations, some male Ospreys bring more fish to their families than others do. Being a good provider seems to hinge on a variety of factors including a bird's experience, its nest location, and, perhaps, its inherent ability or motivation. Our concern here is whether such variation influences the production and survival of eggs and young.

Females given extra food by their mates during courtship might be expected to lay larger eggs and clutches than poorly fed females. Trying to explain variation in clutch and egg sizes among female Ospreys, I became intrigued with this possibility and with the larger question of why Ospreys (and most raptors) engage in courtship feeding when many other birds do not. To find answers, I monitored

Table 7.2. *Characteristics of female Ospreys laying three or four eggs. Data shown are mean values. None of these paired comparisons were significantly different when tested statistically.*

	3 eggs $n=57$ (71%)	4 eggs $n=23$ (29%)
Egg size (ml)	64.8	62.8
Laying date	23 April	22 April
Food intake of female during courtship[a]	72	69
Weight of female at laying (g)	1917	1965
Age of female (yr)	7.8	6.8

[a]Number of bites of fish per hour.
From Poole (1985).

food delivery at Osprey nests and looked for correlations between a female's food consumption during courtship, her weight at laying (an indication of her nutrient reserves), and her clutch size (Poole, 1985). Females laying three eggs ate just as many grams of fish (of roughly the same species) as those laying four (Table 7.2), suggesting that the amount of food a male fed his mate had little influence on her egg production.

Three related findings (Poole, 1985) have supported this conclusion. First, as in other large birds of prey (Newton, 1979), each Osprey egg is a small percentage (about 3%) of the layer's weight, so producing even four eggs demands little more energy than that needed by nonbreeders. Second, even when well fed, female Ospreys gain little weight during courtship (Fig. 7.3), so they probably do not depend on stored body reserves (fat or protein) to help them through laying. Instead, females probably produce their eggs from the food they eat while eggs are developing. They can get away with this because their food is high in fat and protein. Third, a few pairs given supplemental food (fresh fish left at the nest edge once a day) ate the fish readily but laid no more eggs than unfed controls. Fed males, in fact, rarely bothered to hunt, suggesting that both they and their mates had all the food they needed.

In conclusion, egg laying apparently places little nutritional stress on female Ospreys. Courtship feeding in this species has probably evolved for other reasons than to boost egg production. In addition to holding a female at a nest site and keeping her faithful, courtship feeding may stimulate the final development of her eggs. Pairs that start courting early, for example, usually lay early as well (Poole, 1985).

Compared with clutch size, the number of eggs and young that survive in a nest should better indicate a male Osprey's ability to find and deliver food. On a regional scale, as we saw above, broods are usually smallest in areas where food is scarce. Within colonies, however, few studies have compared brood survival at nests where food delivery was also measured. Debra Judge (1983) showed that successful pairs in Baja California (Mexico) brought more food to their nestlings than did pairs which failed to raise young. In contrast, successful females in New England ate no more fish during incubation than females failing to hatch young (Poole, 1984). Most of the latter observations were during fair weather, however. Since nest abandonment often occurs during storms, further observations are needed. Yet even if males do deliver more fish to successful (large) broods, large broods demand more food than small ones, so differences in delivery rates may simply reflect differences in brood size rather than being responsible for those differences. Ideally, a study should start measuring food delivery and consumption soon after hatching, before young are lost to starvation.

Limits to brood size

Many parent Ospreys seem unable to nourish large broods. We see this in the recurrent starvation of third and especially fourth hatched

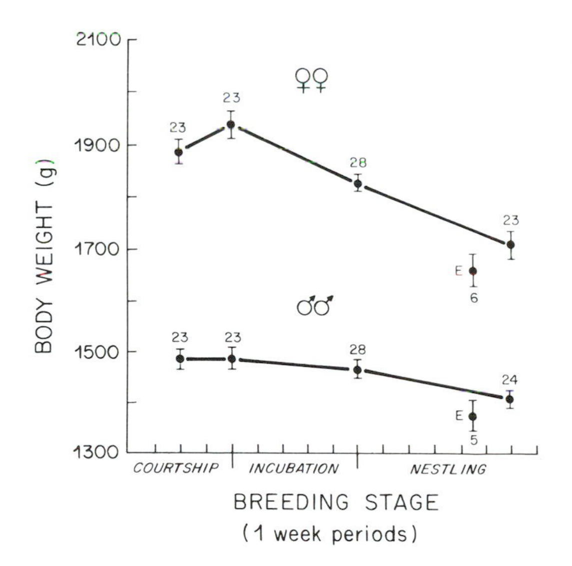

Figure 7.3. Seasonal weight change among Ospreys breeding in southeastern Massachusetts. Means and standard errors are shown. The number of birds weighed at each stage appears above the error bars. Weights designated by 'E' refer to five males and six females that fed enlarged broods. All of these Ospreys were weighed using electronic perch scales, shown in Figure 7.4. From Poole (1984).

Figure 7.4. *Top:* Ospreys like to roost at perches near their nests. This habit has made it easy to weigh breeders using artificial perches containing battery powered scales (Poole & Shoukimas, 1982). *Bottom:* an artificial nesting platform with a perch scale installed. A bird sitting on the perch (a) causes a change in current passing through a strain gauge. The current, proportional to weight, is recorded on a continuously operating chart recorder (b). (Photos: A. Poole.)

young in localities where food is scarce (Appendix 5; Table 6.3). Indeed, watching parent Ospreys, one is apt to conclude that males, the providers, are lazy – perhaps 'unhurried' is a better word. Walk into any Osprey nesting colony, at almost any hour of the day, and you are likely to find most males in residence, not out fishing (Fig. 7.5). The discovery that male Ospreys usually spend only about 15%–35% of the day foraging, and thus 65%–85% of the day preening, eating, and sleeping (hardly energy intensive activities), led Stinson (1978) to conclude that Ospreys reproduce 'sub-optimally' – that they were able to raise more young than they do.

Why do male Ospreys not raise five or seven young each year? What limits their foraging efforts? Males do deliver more food as their chicks age, but this increase levels off when chicks are three or four weeks old (Fig. 6.17). It is not clear if chick demands also level off then or if males run out of energy and simply cannot work any harder.

I tested the foraging limits of male Ospreys by manipulating their brood sizes and then monitoring changes in food delivery and consumption as the same male fed different numbers of young (Poole, 1984). This allowed me to test an individual bird, to see if brood size actively determines, or just passively reflects, foraging rates. Ospreys are ideally suited for such field experiments because neither parents nor nestlings discriminate against foreign chicks, so a chick's access to food is based on size, not kinship. In addition, if transfers are temporary, the birds suffer no harm.

I carried out a series of such manipulations, increasing broods from three to five young, and noted some telling changes (Table 7.3). Delivery of fish did increase after brood enlargement, but not nearly enough to compensate for the added young; there was an overall reduction in food per chick. With less food available for enlarged broods, nestmates fought more, their access to food was rarely equal, and they grew at very different rates.

A second finding, perhaps more revealing, was that female Ospreys, given extra young, ate less food than females with natural broods, while their mates continued to eat about the same amount regardless of brood size (Poole, 1984). Apparently a female faced with hungry, begging nestlings cannot resist feeding them. Males probably avoid such sacrifice by feeding first and away from the nest. Both sexes lost weight tending enlarged broods, but females lost about three times more than their mates (Fig. 7.3). Ospreys feeding natural broods suffer some weight loss during the nestling period,

Table 7.3. *Food delivery and division at Osprey nests with natural and experimentally enlarged broods. Data on the growth and behavior of chicks are included to show how these changed with different food regimes. Each of the natural and enlarged broods was observed a minimum of 30 hours.*

	Natural broods ($n=6$)	Enlarged broods ($n=6$)
No. chicks per nest	3	5
Fish bites/parent/hour[a]	30	22
Fish bites/chick/hour	33	27
Chick aggression[b]	0.5	1.5
Range in growth rate (grams per day)	+25 to +70	−15 to +75

[a] Female parent only.
[b] Scored from 0 (none) through 3 (high intensity).
From Poole (1984).

Figure 7.5. A male Osprey perched near its nest, New England, USA. (Photo: A. Poole.)

but the losses among females with enlarged broods were potentially drastic. If females lost weight at similar rates for 30 days, for example, they would be about 40% lighter than when they started breeding. Such females would undoubtedly desert nests or let most of their young starve well before they became so emaciated. Males may be less tolerant of weight loss because they must hunt daily.

In summary, there are two ultimate limits to brood (and thus clutch) size in Ospreys, these also being the reasons why selection has favored individuals that are 'cautious' parents. First, the quality of young tends to decline with increasing brood size. Compared with small broods, large broods are usually hungrier, more aggressive, and more likely to starve or produce runts. Second, the larger the brood, the more weight parents (especially females) lose and thus the poorer their chances of surviving to breed again. Breeding Ospreys are not really lazy, therefore, but are up against some very real energetic limitations instead (section 5.4; Appendix 3).

7.3 Weather

People living near Ospreys have often observed that they reproduce poorly during cold, wet breeding seasons. Data from south Sweden and from New England tend to support these observations (Odsjö & Sondell, 1976; Poole, 1984). I find dead and runted chicks or cold eggs in Osprey nests more often after prolonged rainstorms than during fair weather. Apparently, some males have trouble catching fish during high winds and heavy rains, causing chicks to starve or incubating females to abandon nests. Ospreys may be unusually susceptible to bad weather because their nests are open to the elements. Although studies have shown that moderate rain and wind often do not affect Osprey foraging success (section 5.4), no one has watched a series of nests during storms. As more foraging data accumulate, storms will no doubt be recognized as a potent selective force with which nesting Ospreys must contend.

Weather may also affect Osprey breeding success more indirectly, by determining how early in the season a female can lay eggs. At northern latitudes, breeding is earliest and most successful in years when lakes thaw early (Wetmore & Gillespie, 1976). It is not clear why early thaws boost reproductive success; Nancy Clum (1986) has suggested that incubating females may be in poor condition after late ice-outs because they lack adequate food to build up reserves. An alternative explanation, considered in the following section, is that late pairs may give up breeding attempts easily, regardless of their condition.

7.4 The effect of laying date on breeding success

During a four-year study of Osprey reproduction in New England, late breeding pairs hatched and fledged far fewer young than their earlier neighbors (Fig. 7.6). This trend was seen in all years, without exception. Similar trends have been found for Ospreys nesting in Florida, Mexico, and North Carolina (Ogden, 1977; Judge, 1983; Hagan, 1986). Ospreys are not unusual in this regard. A reduction in the size and success of late broods is a widespread biological phenomenon, reported for most birds as well as for rabbits, mice, copepods, butterflies, and spiders (Perrins, 1970; Newton, 1979; Toft, Trauger & Murdy, 1984).

Why do late breeders do so poorly? Among Ospreys, reduced clutch size is not the explanation. Although late Ospreys often lay only two eggs (Poole, 1985), this shift is much too small to account for typical seasonal declines in fledging brood sizes. Instead, late pairs reproduce poorly because few of their eggs and chicks survive. Egg loss is especially high, with many late clutches abandoned (Appendix 6).

Neither can the quality of nest sites explain the failure of late breeders. All pairs considered in Figure 7.6, for example, nested on similar artificial platforms. And since all nested within a few miles of each other, all had potentially equal access to food. Moreover, their

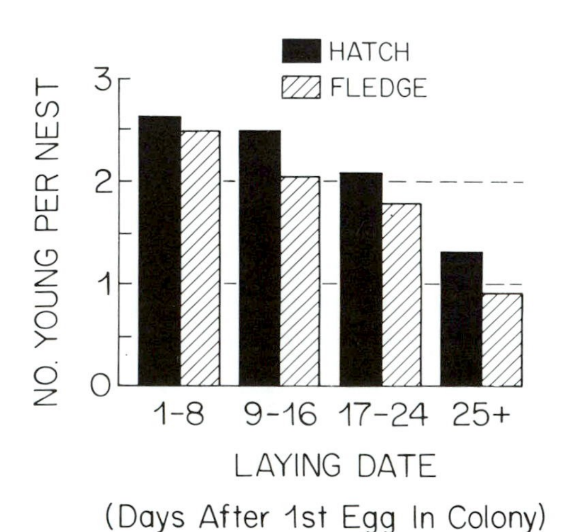

Figure 7.6. Hatching and fledging success of Ospreys that laid at different points in the breeding season. Four years of data (102 clutches) from a New England colony are included here. From Poole (1984).

food supply did not fluctuate significantly during the breeding season (Poole, 1984). Early pairs successfully raised young during the same weeks when many late pairs deserted their eggs, unable to meet the relatively low energetic demands of incubation.

Young Ospreys and those with new mates tend to be the late breeders in most locales (Fig. 6.9), so inexperience could be one reason late breeders do poorly. Preliminary analyses, however, using multivariate statistical techniques, suggest this is not the case. Laying dates alone explain more of the variation in a colony's breeding success than do age and mate retention (Poole, 1984). Larger samples are needed to test these relationships rigorously, but such results hint that even older, experienced pairs raise few young when nesting late. A variety of birds show similar seasonal declines in breeding success within both older and younger age classes (Perrins, 1979; Newton & Marquiss, 1984).

Two other explanations for the high rates of failure among late breeding Ospreys should be considered. First, late breeders may simply be birds in poor condition. Yet in one colony where adults were weighed, early and late pairs weighed about the same, and late pairs still reproduced poorly (Poole, 1985). In addition, entire populations of Ospreys reproduce poorly when late (Wetmore & Gillespie, 1976), suggesting that lack of ability or condition is not the real cause of failure. Second, because young Ospreys that fledge late survive poorly (section 7.5), selection could be expected to favor reduced breeding 'effort' among late pairs. A small clutch is the most obvious change in effort that late Ospreys make, but other facets of reproduction – foraging, food sharing, incubation efficiency, or the tendency to abandon nests – might change too, a possibility that needs investigating. The Osprey is a relatively long-lived, slow breeding species, with high survival among breeders, just the sort of bird that might be expected to slow down (or give up) on one breeding effort in order to ensure its own survival to the next.

7.5 Survival to breeding age

Fledging is no guarantee that a young Osprey will survive to breed. In fact, mortality is severe enough during the first few years of life that less than half the young that leave nests return to build their own (Chapter 8). To study factors promoting survival in young Ospreys, one must band hundreds of nestlings, noting information on each bird (brood size, hatching date, fledging weight), and then identify individuals that return to breed in later years. Survival can then be related to factors in early life. Since young Ospreys often breed near their natal sites (Fig. 8.1), assessing survival of pre-breeding Ospreys is easier than with many other species.

Paul Spitzer and I trapped breeding Ospreys in two regions (Chesapeake Bay and New England, respectively) where nestlings had been intensively banded for many years. We caught and identified nearly one hundred banded breeders between 1980 and 1986. One of our clearest findings was that young hatched early in the season survive to breeding age much better than those hatched late (Table 7.4). These effects were independent of brood size, so even though early broods are the largest, chicks from large broods had no greater or lesser chance of being recovered than those from small broods fledged at about the same time. Early broods may be fed better than others, but we lack a good series of fledging weights or food delivery rates to tell for sure. Young from early broods, however, do have extra weeks in which to gain flight and foraging skills before migration, a luxury which probably boosts their survival.

All of this evidence suggests that some Ospreys are better parents than others, breeding earlier or providing more food for their young, or both. In fact, one does see remarkably strong parent effects – brood mates, or their siblings fledged in other years, tend to survive far better (or worse) than would be expected by chance. During three years of trapping Ospreys in Massachusetts, for example, nearly 50% of the 35 banded breeders I caught had been raised by only *three* pairs in the colony, out of the 18–34 pairs that bred there in each of the years of banding (Fig. 7.7). Almost 55% of these pairs produced no young that were ever trapped. Doubtless some young from these nests survived untrapped elsewhere, but this lopsided recovery pattern was extraordinary. Besides suggesting great differences in parental quality, it indicated that fledging success may be a poor measure of the actual number of new breeders an individual contributes to the generation coming after it.

7.6 Age and mate retention

Like many other animals, Ospreys become more successful breeders as they gain age and experience, at least until old age sets in (Fig. 7.8). Likewise, Ospreys retaining a mate from the previous season reproduce considerably better than those with a new mate (Table

Table 7.4. *Survival of young Ospreys between fledging and first breeding in relation to how early in the season the young hatched. A chick's hatching date was determined relative to average hatching dates for its region. Data from Massachusetts and Maryland.*

	Hatching date		
	Early	Middle	Late
No. of young banded	306	800	200
No. recovered locally as breeders	32	55	7
% recovered as breeders	10.5	6.9	3.5

A. Poole & P. Spitzer (unpublished).

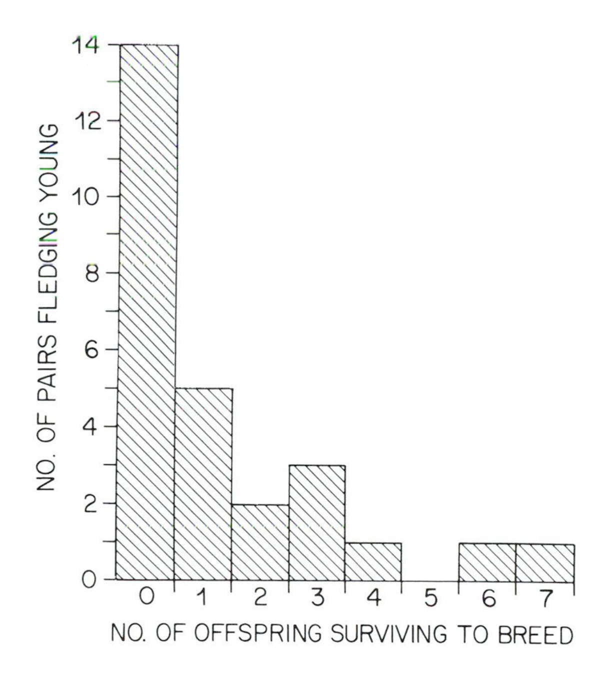

Figure 7.7. Production of fledgling Ospreys that survived and bred locally in Westport, Massachusetts, 1979–1982. Most pairs, it can be seen, fledged no surviving offspring, while 22% of the pairs produced 74% of this area's surviving young. It is unlikely that many surviving young bred outside this area (see Figure 8.1). From Poole (1984 & unpublished).

7.5). These relationships are among the better known facts of Osprey biology, largely because a few populations have been thoroughly color–banded, creating clusters of breeding adults of known age and identity.

Intuitively, one expects older birds with familiar mates to be better parents, but the reasons for this success are hard to pinpoint. Comparing the productivity of six to eight-year-old Maryland Ospreys (females) with that of the same age Massachusetts Ospreys (Fig. 7.8), for example, one sees that age alone is no guarantee of success. Due to scarcity of nest sites, most six to eight-year-old Maryland Ospreys are breeding for only the first or second time. In Massachusetts, birds that old are usually veterans of three or four breeding seasons. Breeding experience, therefore, rather than age *per se*, is the real advantage to Ospreys raising young.

Because male Ospreys provide all the food for their families, their experience is especially important to the reproductive success of a pair. Pairs in which both members lack previous experience produce the fewest young, but, if a male has bred even once before, he is likely to do well the next time, no matter how inexperienced his mate (Table 7.5). The reverse does not hold, however. Experienced females with inexperienced mates usually do poorly (Table 7.5). It is not clear why a year's experience is such an advantage to the male parent, but increased familiarity with local hunting grounds is an obvious possibility.

This chapter has explored the key factors controlling Osprey breeding rates, and thus the number of new breeders potentially recruited to a population. But populations do not grow indefinitely. There are limits to the number of Ospreys that any one region can support. In the chapter that follows, we see how Osprey populations are naturally regulated, why the world has only about 30 000 pairs of Ospreys instead of 300 000 or 5 000 000.

Table 7.5. *Osprey reproductive success in relation to the number of years of breeding experience possessed by the male and female in each pair. To facilitate comparisons among years, hatch and fledge figures are shown as the number of young above (+) or below (−) yearly means for the colony. Data were gathered from banded individuals breeding in Massachusetts, 1981–1985.*

	No. years of breeding experience Male:female				
	0:0	0:1+	1+:0	1:1	2+:2+
No. of pairs studied	27	7	15	17	76
No. young hatched	−0.86	−0.50	+0.08	+0.29	+0.37
No. young fledged	−0.74	−0.64	+0.31	+0.34	+0.48

A. Poole (1984 & unpublished.)

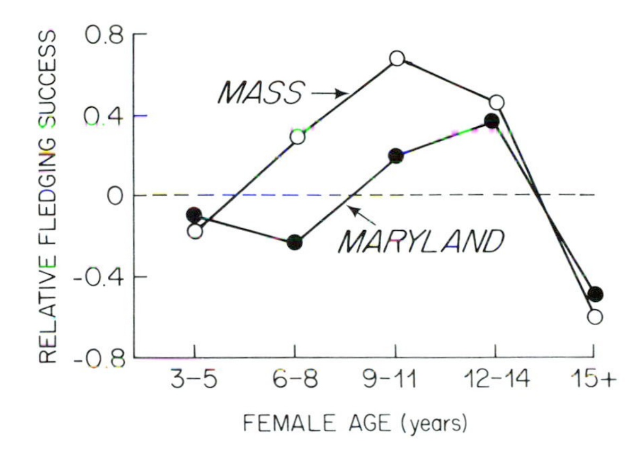

Figure 7.8. Osprey breeding success in relation to the age of the female parent. (Males show similar trends). To facilitate comparisons among areas and years, success for each age group was calculated relative to regional averages. Data from two populations are combined here: Westport, Massachusetts ($N=87$; Poole, 1984 & unpublished) and Chesapeake Bay ($N=104$; P. Spitzer, unpublished).

8 Population Regulation

Effective conservation of raptors ultimately depends on a thorough understanding of what regulates their numbers . . .

(Ian Newton, 1979)

Mathematicians view the dynamics of animal populations as an equation with multiple parameters. Field biologists view such equations with suspicion, however, because they know that finding accurate values for population parameters is often difficult. A simplistic approach suggests that subtracting deaths from births will yield each population's rate of change, but the process is actually more involved. To predict how Osprey breeding numbers are likely to change in any one region, for instance, one must know, in addition to birth and death rates, at least the following: how far the birds are likely to disperse from natal and breeding areas, at what age they start breeding, and what numbers their habitat can support. This chapter examines such parameters for Osprey populations and, comparing findings for different regions, considers how these parameters vary with habitat.

Probably the toughest and most critical question to answer about Osprey populations is what limits their numbers. Populations obviously cannot grow indefinitely. Newton (1979) has summarized evidence that shortages of either food or nest sites are likely to limit most species of raptors in most areas. For Ospreys, such evidence is still preliminary, although this chapter argues that lack of nest sites has been the key constraint for most populations studied so far.

The Osprey's remarkable attraction for artificial nest sites has provided much of the evidence for this hypothesis. Widespread efforts to build nesting platforms for these birds, along with the recent proliferation of nests on power poles and channel markers (Chapters 3, 10 and 11), have constituted large scale (albeit unplanned) experiments in population regulation. Similar experiments manipulating the Osprey's food supply are obviously more difficult to undertake, perhaps biasing our perspective. Nevertheless, Ospreys differ from most other birds of prey in that they do not defend feeding territories (Chapter 5). In any one locale, therefore, many breeders can share the same fishing grounds, but each pair must compete for its own nest site.

By necessity, this chapter is concerned exclusively with migratory populations of Ospreys. The dynamics of resident populations, dependent on warmer subtropical habitats, are sure to differ from those of migrants, but so far no one has undertaken the detailed, long-term studies needed to illuminate these facets of the resident Osprey's ecology.

8.1 Dispersal

There is more to understanding the ebb and flow of Osprey numbers than simply recording births and deaths. Birds are such mobile creatures that the potential for exchange among populations, especially migratory ones, is high. To decipher the dynamics of an Osprey population, therefore, one must know to what extent its breeders are immigrants, raised and nourished elsewhere.

Assessing dispersal requires identifying individuals and their origins, generally by leg bands. In two long-term studies, one in New England and one in Sweden, researchers were able to measure natal dispersal – the distance between fledging and breeding sites – for over 180 Ospreys banded as nestlings. Most of these young, after having migrated to distant subtropical wintering grounds, returned remarkably close to where they had fledged (Fig. 8.1). Apparently the memory of home is well defined and stored in an Osprey's neural circuitry, just as it is for so many other long distance migrants.

Swedish Ospreys, it was found, nested farther from natal sites than Ospreys in New England did (Fig. 8.1). There is no reason to suspect that the Swedish birds were less efficient at finding home territory, so their tendency to disperse longer distances probably reflects difficulty finding food or unoccupied nesting sites. Food is unlikely to be the problem, however; brood sizes, a reflection of food availability, have been equally large in Sweden and New England in recent decades (1.8–2.2 young per successful nest: Odsjö & Sondell, 1976; Poole, 1984). Nesting sites, on the other hand, are scarce in Sweden, but abundant along the southern New England coast where artificial nest sites have proliferated in recent decades (Fig. 8.2; Chapters 10 and 11). Assuming new breeders return first to their natal sites, they have only to disperse a short distance in New England before finding a vacant nest site. Longer dispersal distances, such as those seen among Sweden's tree-nesting Ospreys, are probably the norm for the species.

Compared with females, male Ospreys are more likely to nest near their natal sites (philopatry), a bias found both in Sweden and in the United States – regardless of breeding density and resources, that is

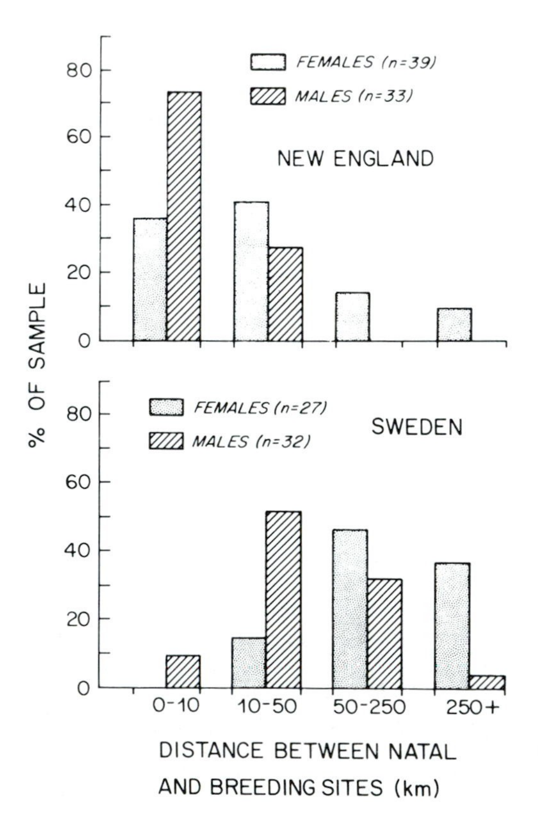

Figure 8.1. Distances between fledging and breeding sites of Ospreys born in southern New England and in Sweden. The New England data (Spitzer, Poole, & Scheibel, 1983), were gathered between 1970 and 1978 by trapping breeders banded as nestlings. The Swedish data (Österlöf 1977 & unpublished), were gathered prior to 1975 from banded Ospreys of breeding age found dead during the nesting season.

Figure 8.2. Typical nest sites of Ospreys in Sweden (top) and in New England (bottom) during the years when natal dispersal distances were being studied in these populations. (Photos: J. Sondell (top) & A. Poole (bottom.)

(Fig. 8.1). In New England, such male philopatry was extreme. Only two out of every 10 males settled more than 10 kilometers from home and none dispersed farther than 50 kilometers. Some degree of philopatry is typical for male birds of nearly all species (Greenwood, 1980). There are two reasons why selection might be expected to favor male Ospreys that return to natal areas. Familiarity with a locale would help males find nest sites, the critical resource for attracting mates. Furthermore, where a male managed to fledge and survive, its offspring should do well also, given that ecological changes come slowly to Osprey habitat. Females, on the other hand, might be expected to disperse farther because their only imperative is to locate a free male with a nest, and wandering would enhance that search.

It is not yet known exactly how young Ospreys disperse in their search for a breeding site, whether survivors return first to their natal region and then wander away, if need be, or whether some stop off during migration without ever returning home. Scandinavian Ospreys, probably migrants, recolonized Scotland during the 1950s (Chapter 11), suggesting that the trip home can be short–circuited. A more critical and perplexing question concerns the costs of dispersal. Besides having to delay breeding, individuals dispersing into new territory probably suffer high mortality, although we lack reliable data to support this contention.

Several implications of the Osprey's relatively short natal dispersal distances are clear, however. Such fidelity means that local reproduction will usually determine a population's stability. Indeed, this helps to explain why, without the bolstering effects of immigration, many US Osprey populations fell so precipitously during years when pesticides cut breeding rates to almost nil (Chapters 9 and 11). An unwillingness to disperse far also means that Ospreys are slow to colonize new areas. They took over 40 years, after all, to recolonize Scotland from Sweden (only 600 kilometers away), even though Scandinavian migrants pass through Scotland every spring and fall (Chapter 11).

Once a nesting site has been chosen, an Osprey is generally faithful to it. Among color-banded breeders in New England, only 3% changed nesting sites over a 10-year period and no bird nested more than 18 kilometers from its old site (Fig. 8.3; Spitzer *et al.*, 1983; Poole, 1984). Ospreys breeding along eastern Chesapeake Bay show similar trends (P. Spitzer, unpublished). In both these regions, however, most Ospreys nest at stable artificial sites, structures that generally last for years. Where Ospreys nest in dead trees and other more precarious natural sites (Table 8.4), year-to-year fidelity to breeding sites is undoubtedly much lower.

8.2 Mortality

Death, for a population biologist, is the debit side of a ledger. As G. Evelyn Hutchinson (1978), the dean of population biologists, has written: 'If we want to think intelligently about how population growth is controlled by natality, we find paradoxically that it is best to start by thinking about death.' Accurately estimating mortality rates, however, has proven as difficult with Ospreys as with most other wild birds. Basing estimates on bands recovered from dead birds is generally unsatisfactory because there are so many biases involved in collecting the data, especially among migratory populations (Chapter 4).

Paul Spitzer (1980) sought one way around this problem by

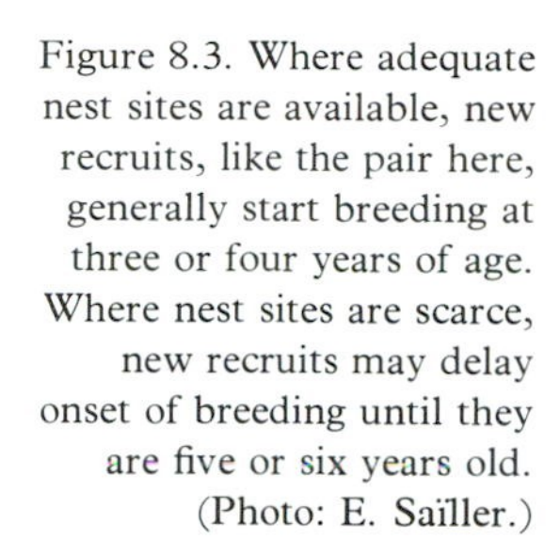

Figure 8.3. Where adequate nest sites are available, new recruits, like the pair here, generally start breeding at three or four years of age. Where nest sites are scarce, new recruits may delay onset of breeding until they are five or six years old. (Photo: E. Saïller.)

estimating survivorship for a stable Osprey population – one in which the breeding rates needed to balance mortality were known. In such cases, a simple equation links births and deaths (Henny & Wight, 1969). Spitzer thus estimated that 85% of the adults (more than two years old) he studied survived each year. This estimate assumed a constant rate of mortality after the first year of life, which is probably not the case for any Osprey population, as shown below.

Another approach to the problem of estimating mortality, one that accounts for variation with age, is to follow the year-to-year persistence of individuals in a breeding population. Because most Ospreys are faithful to breeding sites, one can reasonably assume that any established breeder failing to return to its nesting territory (or nearby) at the start of the season has died. Determining mortality for a color-banded population of Ospreys is thus a routine matter. As breeders perch near their nests or fly overhead, a quick check with binoculars or telescope will usually reveal their identifying colored leg bands. Thus one knows which of the past year's birds have survived and returned to that locality.

Years of intensive banding are starting to pay off. There now exist preliminary, but direct, measures of adult mortality for two US Osprey populations (Table 8.1). Annual mortality of these birds has ranged from about 10%–17% in recent years – close to Spitzer's earlier estimates. In both populations, Ospreys from 10 to 13 years old were more likely to die than others. Ospreys produce more young at these ages than at other periods in their lives (Fig. 7.8), so we may be glimpsing some of the costs of heightened reproductive effort here. These overall death rates are lower than those of small raptors and higher than those of large eagles (Newton, 1979), but quite similar to those of small seabirds like terns (DiCostanzo, 1980). Thus mortality rates vary predictably with body size among different birds of prey, but not necessarily among different orders of birds.

For any wild animal, youth and inexperience carry great risks – loss of direction, starvation, accidents, predation. Ospreys are probably no exception. Yet mortality of pre-breeders has been more difficult to determine than that of breeders because the young birds are elusive, not yet attached to nest sites and usually scattered throughout distant wintering grounds. Using band recoveries, Henny & Wight (1969) estimated that only about 50% of all Osprey fledglings survive their first year of life. Spitzer (1980) refined this estimate to about 60%, given a stable population with adult survival running 85% per year. Both these estimates were really guesses; actual values are sure to vary among populations.

Table 8.1. *Mean annual mortality of Ospreys of different ages in two eastern US breeding populations, 1982–1986. Mortality is shown here as the percentage of each age class that failed to return to the breeding colony each year and thus were presumed dead. ND means insufficient data were available to calculate mortality for an age class.*

	Age class (years)				
Region	3–5	6–9	10–13	14–17	18 +
Massachusetts					
no. of breeders	54	74	41	10	5
% dead	11.1	9.5	17.1	ND	ND
Chesapeake Bay					
no. of breeders	14	113	135	43	11
% dead	ND	10	15	12	ND

A. Poole & P. Spitzer (unpublished).

In the meantime, given what we know already about Osprey mortality, it is clear that each cohort is quickly whittled down to a handful of survivors. On average, out of one hundred fledged young, 37 will be alive four years after fledging, 17 eight years after, and only six to eight 12 years after. In three instances, an Osprey 24–25 years old has been found breeding (Spitzer, 1980), the greatest longevity recorded for the species and truly remarkable survival considering the figures above.

8.3 Age at first breeding

The younger that birds are when they start breeding, the sooner they start contributing new members to a population and the faster that population can be expected to grow. But reproducing successfully requires skills gained only with experience, so among long-lived birds like Ospreys there is usually a lag of several years between fledging and first breeding.

By checking for banded recruits in southern New York and New England, Spitzer (1980) found that, of the 20 young Ospreys he identified as nesting and producing eggs for the first time, 50% were three-year-olds, 30% four-year-olds, and 20% five-year-olds, making the mean age at first breeding about 3.7 years in this population. Expanding this small sample to include 40 more new breeders in this region has revealed a similar pattern: most of these Ospreys start breeding as three- or four-year-olds, with a mean age at first breeding of 3.6 years (Poole, 1984).

Between 1970 and 1985, when these data were gathered, Ospreys in this region nested at low density in relation to available resources. Thus most new breeders had little difficulty finding food or nesting sites. As noted in Figure 8.2, vacant platforms and other artificial nesting sites were plentiful in this region. Young Ospreys returning to breed usually settled in quickly after arrival (Poole, 1985). Their only real task was to choose among a variety of prime nesting locations, undoubtedly a key reason for not dispersing far (Fig. 8.1).

Compare these birds with those nesting along the eastern shore of Chesapeake Bay, where most Ospreys do not start breeding until they are five to seven years old (mean age = 5.7 years; P. Spitzer, unpublished). Only tall trees or overwater structures like duck blinds and channel markers provide predator-proof nesting sites for these birds. Most such locations, at least those with room to support a nest, already have Ospreys in residence. New recruits along Chesapeake shores are thus faced with the formidable task of displacing an older, established nesting bird or pioneering some marginal site that other Ospreys have ignored, probably for good reason.

Many of these Chesapeake young must therefore delay breeding, waiting until they find a single bird with an established nest, perhaps an older breeder whose mate has died. In fact, young Chesapeake Ospreys do pair often with older birds, while young New England Ospreys show a significant tendency to pair with individuals close to their own age (Table 6.1). Thus age at first breeding, like natal dispersal distance, varies not only among individual Ospreys but among populations, apparently in relation to the availability of nest sites and, perhaps, other resources. People expecting such parameters to be constant for all Osprey populations are forgetting that habitat shapes most aspects of a species's life history and that the Osprey's breeding habitat is anything but uniform (Chapter 3).

8.4 Population stability: breeding rates and annual recruitment

How many young per nest does an Osprey population need to produce in order to balance mortality and stabilize its numbers? This is a key question, one asked often of those who study Ospreys. Here again, the answer is sure to vary among regions, although too few populations have been studied to know this for a fact. What is known, however, is that different methods of calculating production requirements will yield different answers. Henny & Wight (1969), using band recovery data to estimate Osprey mortality in the eastern United States, calculated that an annual productivity of 0.95–1.30 young per active nest would balance deaths.

Given the biases inherent in band recoveries, greater accuracy can be achieved by measuring production requirements directly, with live birds in the field rather than with dead birds reported by the public. Paul Spitzer (1980) used this direct approach in his 12–year study of Osprey population dynamics in southern New England and nearby eastern Long Island. Thanks to a broad network of competent helpers, he was able to track breeding rates and population change simultaneously in this coastal population (Fig. 8.4), and to link the two parameters afterward. A number of unique circumstances made this study possible. For one thing, Ospreys here were especially faithful to natal sites (Fig. 8.1) and no other population bred nearby, so the birds were relatively isolated reproductively. In addition, most nests were clustered and accessible, so breeding rates and numbers could be monitored accurately as the population grew. Spitzer also knew age at first breeding for these birds, which helped in making accurate estimates of annual recruitment. Finally, breeding rates more than doubled during the study period, due to declining contamination of eggs, allowing reproductive variation to be correlated with changes in breeding numbers.

Spitzer (1980) calculated that about 0.80 young per active nest was the reproductive break even point, the breeding rate needed for stability, in the population he studied (Fig. 8.4). The rate at which new breeders are recruited to an Osprey population in any one year is a function of the age at which these birds start breeding, the population's breeding rates in the years supplying the new breeders, and the survival rates between fledging and first breeding.

Assuming that young fledged in different years have roughly equal chances of survival, one can calculate a figure, weighted to account for the differing percentages of recruits returning to breed at different ages, for each year's expected level of recruitment. This Adjusted Recruitment Productivity (ARP) provides one measure of the potential pulse of new breeders arriving each year in a population. By plotting annual change in breeding numbers against annual ARP (Fig. 8.5), Spitzer (1980) showed that his recovering study population first stabilized when ARP was about 0.80 young per active nest, suggesting this was the pivotal figure. When ARP was greater than 0.80, breeding numbers rose, continually swelled by new recruits.

Such extraordinary rates of population growth depend, of course, on a plentiful supply of food and nest sites. The southern New England population, reduced to about 15% of its 1930–1940 numbers when Spitzer started his study, bred at very low density in

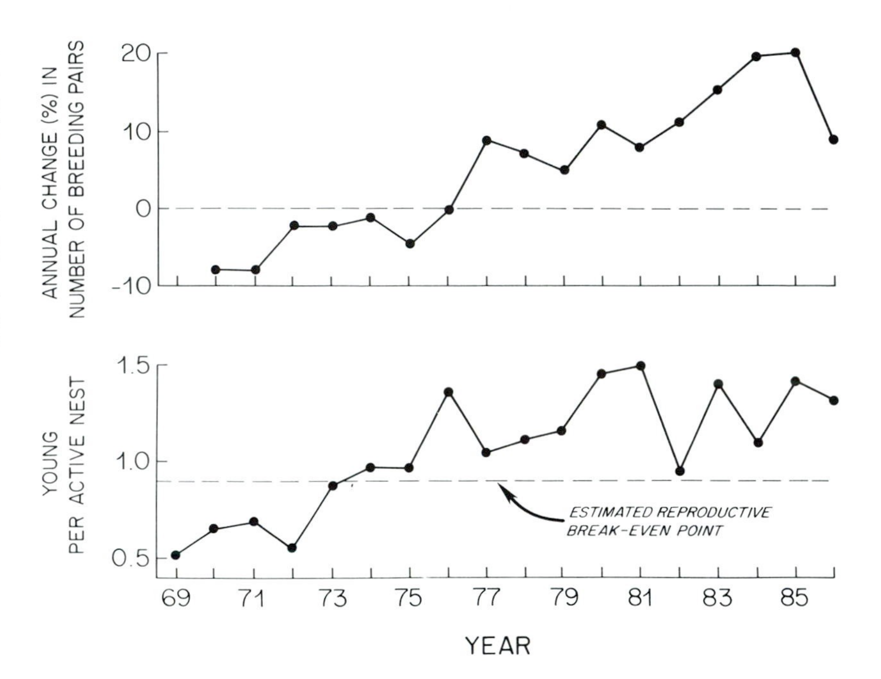

Figure 8.4. Yearly changes in Osprey reproductive output and breeding numbers along the coasts of southern New England and eastern Long Island, NY, 1970–1985. The growth in productivity reflects declining levels of DDT residues in eggs, with accompanying increases in egg viability (Chapters 9 and 11). From Spitzer *et al.* (1983 and unpublished).

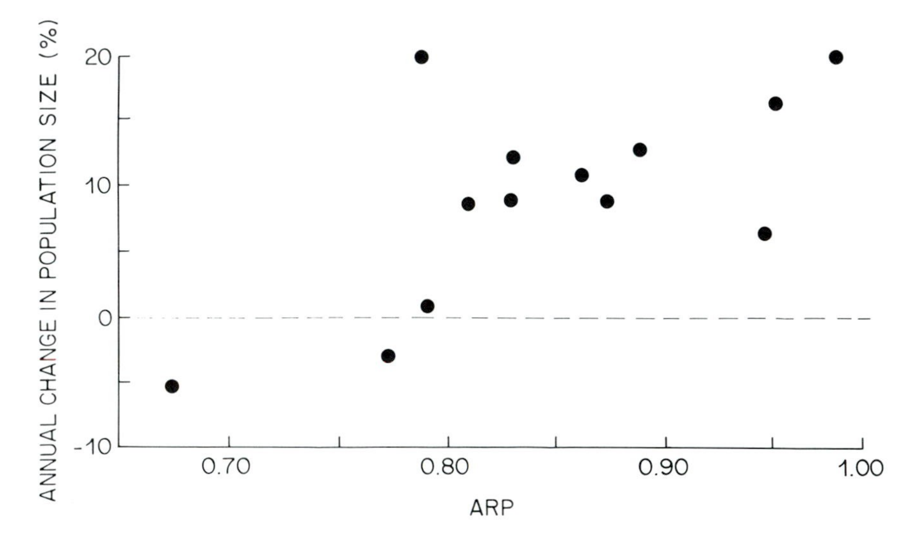

Figure 8.5. Relationship between Adjusted Recruitment Productivity (ARP) and annual change in the size of the Osprey population described in Fig. 8.4. In other populations, the reproductive levels needed for stability are likely to be higher (see Table 8.2). From Spitzer *et al.* (1983 and unpublished).

Table 8.2. *Estimated relationships among Osprey population parameters. Breeding rates required for population stability were estimated in relation to parameters,* A *and* S, *for coastal Ospreys breeding in southern New York and New England for which Adjusted Recruitment Productivity (ARP; see text) was found to be 0.80. For these birds,* $R = 0.80/S^{(A-3.7)}$.

Mean age at first breeding (years) (*A*)	Annual survival after age 3 (*S*)	Breeding rate needed for stable population (young/active nest) (*R*)
3.7	85%	0.80
4.7	85%	0.94
4.7	80%	1.00
5.7	85%	1.11
5.7	80%	1.25
6.7	85%	1.30
6.7	80%	1.56

After P. Spitzer, 1980 & unpublished.

relation to these critical resources during the 1970s and 1980s (Chapter 11). But how must breeding rates change to stabilize populations whose food and nest sites are scarce? Such constraints, after all, are probably the norm for most Ospreys. Consider recruits that delay breeding for two or three years past sexual maturity, as those along Chesapeake Bay do. If 10%–15% of each cohort dies annually, then fewer pre-breeders would be alive to join a population at age six than at age three, and higher breeding rates must offset this loss (Table 8.2). A breeding rate of 1.15 young per active nest in a region like Chesapeake Bay where Ospreys breed at a mean age of 5.7 years would thus be equivalent to 0.80 young per active nest in a region like southern New England where Ospreys breed at a mean age of 3.7.

Such calculations assume, of course, that an Osprey's reproductive lifespan is the same no matter when it starts breeding. This may not be the case. By delaying breeding, Ospreys may actually live longer, although there is no proof of this.

8.5 Nest sites: direct effects on population trends

So far, we have considered how nest site availability might affect the stability of an Osprey population by influencing the age at which

Table 8.3. *Breeding success of Ospreys nesting in trees (natural sites) and on platforms, towers, or channel markers (artificial sites) in regions of the United States where both types of nests exist together in similar habitat.* n = *the number of nests monitored.*

	Mean no. of young per active nest at:		
Region	Artificial sites (*n*)	Natural sites (*n*)	Study
Michigan	1.2 (234)	0.6 (126)	Postupalsky, 1978
Florida	1.5 (49)	0.7 (45)	Westall, 1983
Maryland	1.04 (164)	0.7 (286)	Reese, 1977
New York	1.23 (285)	0.92 (399)	[a]
Idaho	2.2 (18)	1.3 (109)	VanDaele, 1982

[a] M. Scheibel, NY State, Dept. of Environmental Conservation.

Table 8.4 *Stability of Osprey nests built in trees in various regions of North America. Most nests that were lost blew down in windstorms.*

Region	% nests lost per year	Study
Florida	50–70	Poole, 1984
New York	30–40	Poole, 1984
California	30	Airola & Shubert, 1981
Mexico	18–44	[a]
Montana	10–15	Grover, 1983
Maryland	10	Reese, 1977
Ontario	5	Grier *et al.*, 1977

[a] F. & F. Hammerstrom, unpublished.

individuals first breed and their fidelity to natal and breeding areas. Here, we consider how both the quality and availability of nest sites can influence Osprey breeding rates and density, two key aspects of population regulation. Breeding rates, after all, drive population growth, while lack of nests can slow this growth effectively, capping the numbers a region can support.

At the population level, one of the most obvious effects of providing Ospreys with stable, artificial nest sites has been increased breeding success. In a variety of habitats, pairs nesting on artificial

Table 8.5. *Increase in the number of Ospreys breeding along the coast of southern New England and New York State, 1976–1986, in relation to nest site availability. Regions with the most nest sites available to Ospreys recruited the most new pairs. Data were collected by state wildlife agencies in each region.*

Region	% of population nesting on artificial sites	Growth of population (%)
Massachusetts	96	395
N Long Island	92	480
Connecticut	90	290
Gardiner's Is.	25	78
Shelter Is.[a]	0	60

[a] Mashomack Preserve.

sites have consistently raised more young per breeding attempt than neighbors with nests in trees (Table 8.3). Fledging brood sizes in the two types of nests have been roughly similar, so it is reduced nest failure – total loss of eggs and young – that sets the artificial sites apart. Predation could be a factor here, because tree nests are often vulnerable to climbing predators (Chapters 3 and 9), but lack of stability is probably the main cause of failure. While nests on artificial sites rarely succumb to winds, storms often blow down tree nests (Table 8.4). Ospreys nesting in dead trees (snags) or in short, flexible live trees like mangroves are especially vulnerable to windstorms, while nests in tall, stable forest trees usually fare better.

One is reminded here of John Muir, the celebrated North American naturalist, who was said to have ridden out a hurricane sitting in an eagle's nest. He wanted the feel of the storm, apparently, from an eagle's perception of the elements in all their fury. Had Muir chosen an Osprey nest as his storm-tossed vantage point, he might never have lived to write about it. I have sat (briefly and during fair weather) in a few empty Osprey nests, but most will not support a person's weight.

Knowing that Ospreys are more likely to fledge young at artificial nest sites than at natural ones, and that most young return to breed near their natal sites, it is no surprise that Osprey breeding numbers have increased most dramatically in regions like New England where artificial nest sites are abundant (Table 8.5). Admittedly, New England populations were primed for growth during the 1970s and

1980s because their numbers had crashed during the two previous decades (Chapter 11). But strong population growth stimulated by extra nest sites has been noted also in the western and southeastern United States where no crash had occurred (Chapter 3). In North Carolina, newly flooded swamps, with overwater nesting trees that are really the equivalent of artificial sites in terms of safety and attraction to Ospreys, have gained new recruits much faster than surrounding upland areas (Henny & Noltemeier, 1975; Hagan, 1984).

One might suspect that swamp trees and nesting platforms boost Osprey numbers because, as prime nesting locations, they drain breeders from other locales. But growth of such colonies has usually involved more than the simple rearrangement of a region's nesting pairs. In southern New York and New England, for instance, the number of tree nesters remained relatively stable between 1976 and 1986, while regional numbers soared due to new recruits at platforms (Table 8.5). During this one decade, in fact, platform nesters jumped from about 25% of the population to over 65%. Similar trends were seen in Wisconsin between 1975 and 1984 (Eckstein & Vanderschaegen, 1988). Thus nest site quality and availability boost local numbers mainly by enhancing breeding rates and by retaining new recruits when they return.

Where food is scarce, where platforms are poorly sited, or where natural nests are safe and abundant, however, platforms may fail to recruit new breeders quickly. The Finnish Osprey population, for example, grew little between 1971 and 1985, even though Finns built over 200 nesting platforms during those years (Saurola, 1978 and 1986). This population has shown consistently strong breeding

Figure 8.6. Ospreys breed at their highest densities on islands free of mammalian predators. Such locations allow safe nesting in low trees and even on the ground, as shown here. (Photo: H.H. Cleaves, Archives, Staten Island Institute of Arts and Sciences.)

Table 8.6. *The mean distance between Osprey nests in four regions of eastern North America where differing types of nest supports are available.* n *is the number of nests sampled. Only active nests are considered here. In most other regions, Osprey nests are farther apart than this.*

Location	Nest type (n)	Nearest neighbor distance (m)	Study
New York	Upland trees (17)	410	Poole, 1984
N Carolina	Swamp trees (35)	170	Hagan, 1984
Massachusetts	Platforms (33)	140	Poole, 1984
Nova Scotia	Island ground (9)	50	Greene, 1987

success in recent decades, so it seems unlikely that lack of food has put the cap on breeding numbers. Poorly sited platforms, advanced age at first breeding, or high adult mortality rates might each be a likely explanation for the lack of growth. Whatever the case, the contrast between the explosive growth of Osprey numbers along the New England coast and the stability of Ospreys dependent on forest lakes in Finland deserves further study.

Breeding Ospreys have always concentrated where nest sites were safest and most available. Throughout the world, islands free of mammalian predators have been favored nesting locations because such habitat allows safe nesting in low trees and even on the ground (Fig. 8.6), greatly expanding a pair's choice of potential nest sites. Swamps, as we have seen, also provide safe, abundant nest sites (Chapter 3). Nearby mainland environments rarely support the nesting aggregations that swamps and islands do, even though locally all birds have access to the same foraging grounds.

The natural dispersion of Osprey nests, therefore, suggests that breeding density is controlled to a great extent by the availability of stable, predator-free nest sites. One finds experimental support for this hypothesis in regions with clusters of artificial nesting platforms. In many such cases, platforms have let breeders nest closer together than was possible in most natural habitats. In New England, the density of occupied platforms is generally much higher than nest density in surrounding forested areas. Only swamps and islands allow nest densities comparable to those of platform colonies (Table 8.6). In addition, increased nest site availability in the form of platforms and other artificial sites has allowed Ospreys to invade new habitat such as urban areas and shallow lakes and marshes (Chapters 3 and 11). In many different areas, therefore, Osprey breeding numbers are restricted by lack of nest sites.

In summary, Osprey populations tend to grow most quickly where nest sites are safest and most available. Reproductive success is generally highest in such locations, producing extra recruits, most of which settle near natal sites. In addition, safe, stable nest sites seldom blow down, enhancing population stability. And where safe nest sites are most available, pairs nest close together and start breeding at an early age, reducing the breeding rates needed to stabilize a population. Nest site availability can thus be expected to regulate Osprey populations in a density-dependent manner: as a population grows and local nest sites become scarce, new recruits disperse farther, delay breeding, and opt for marginal nest sites, all of which cut the number of new breeders joining that population.

In the following chapter, we see how certain Osprey populations have been held below the carrying capacity of their habitat by human ignorance, greed, and persecution. In addition, Chapter 9 considers threats that Ospreys may face in the future as the human population doubles in the next 20–30 years.

9 THREATS

The first rule of intelligent tinkering is to save all the parts.

Aldo Leopold (1949)

All birds, even birds of prey, face a variety of natural threats in their lives. Such threats, in fact, are key forces shaping the form, capacities, and behavior of species over evolutionary time. Before industrial man, Ospreys were widespread and abundant, apparently well adapted to whatever dangers they encountered. Predators or climatic change may have eliminated local populations on occasion, but overall the species thrived. Industrialization, however, gave humans the power to alter ecosystems on a grand scale and at unprecedented speed. Guns, chain saws, toxic chemicals, synthetic fish nets: each brought potential for greater ecological harm, over a shorter period of time, than anything the Osprey had faced before.

Despite such threats, Ospreys have survived this onslaught remarkably well, far better than most other large birds of prey. With modest help, they should continue to do so. This chapter examines both natural threats to Ospreys, such as predators, and recent threats from the activities of humans. Also considered here and in the two chapters that follow are the particular characteristics of the Osprey that have allowed it to emerge from decades of persecution, habitat alteration, and food contamination with significant breeding populations still intact. What is it about Ospreys that suggests their chances of surviving into the twenty-first century are at least as good as ours?

It was as an indicator species that Ospreys gained much of their fame. Their failing reproductive health and numbers were warnings of contamination in the waters where they fished. Indeed, most of what we now know about Ospreys was learned between 1965 and 1980, a period when Osprey studies proliferated due to concern about a few contaminated populations. Yet Ospreys are indicators of more than toxic substances. Historical fluctuations in Osprey breeding success have often reflected other sorts of ecological change – loss of forests or fisheries, for example. Furthermore, the extent to which these birds have been persecuted at different times in different countries shows just how fickle attitudes are towards wildlife. The Osprey's welfare, therefore, indicates much about the material environment and human cultural milieu in which it lives.

Some people, hearing of threats to Ospreys, despair, seeing each individual death as a step toward the demise of the species. Yet different threats have very different consequences for populations. Ultimately, of course, it is the long-term stability of breeding numbers that measures the severity of a threat. Population consequences must be kept foremost in mind when considering the real problems Ospreys face.

9.1 Predators

Perhaps not surprisingly, records of Ospreys killed by other animals are rare. Even experienced naturalists seldom see wild animals die of natural causes. Crocodiles have been known to snatch a few unwary Ospreys roosting on mudbanks (Prevost, 1982), but circumstantial evidence suggests that only owls kill adult Ospreys with any regularity. I found seven Osprey carcasses, all incubating females, all with puncture (talon) wounds in their bodies, and all partially eaten, during three years of nest surveys in a New England breeding colony of about 40 pairs. Great Horned Owls (*Bubo virginianus*) nested near this colony in both years and actually laid eggs in an Osprey nest one year. Large, powerful, nocturnal predators (they can kill full grown house cats), these owls probably find incubating Ospreys easy targets. In fact, there is really no other predator in this region from which an adult Osprey could not readily escape or defend itself, even at night. Ospreys can fly at night, so the owls must surprise them at the nest. Although never witnessed, therefore, such killings seem quite possible and suggest that other large owls in other regions may also threaten nesting Ospreys.

Osprey eggs and chicks are obviously more vulnerable to predation than adults. Although Ospreys will defend their nests, often agressively (section 6.7), predators sometimes succeed anyway. Owls have been suspected of killing nestling Ospreys in Scandinavia (Saurola, 1986) and elsewhere, but Raccoons, agile tree-climbing predators, may be a more serious problem in the United States. North American Raccoon populations have exploded during the past few decades, thanks to this animal's fondness for suburban garbage. Along with other mammalian predators in other regions, Raccoons are quick to take advantage of Ospreys that fail to find inaccessible nest sites. I often find evidence of Raccoon predation – broken eggshells below nests or claw marks and tufts of hair on nest supports. In one US Osprey colony, 30%–50% of the nests fell prey annually to Raccoons before metal guards (Appendix 9) were installed at nest trees.

Although Ospreys do lose eggs and chicks to climbing predators, they are remarkably adept at recognizing nest sites that these mammals cannot reach. Island nesting colonies attest to this ability, as does the Osprey's strong attraction to nest sites over water. In the colony I study, shoreline nests are robbed more often than island nests, apparently because Raccoons cannot survive year–round on small islands and seem reluctant to swim to them.

9.2 Shooting and trapping

Along with most other large birds of prey, Ospreys have been killed by a variety of human predators: hunters, gamekeepers, overzealous collectors, and bored marksmen, especially since the invention of the shotgun. Far too many Ospreys are now stuffed, collecting dust on mantels, or reduced to a set of claws in some forgotten drawer. Sad and regrettable as this killing has been, its actual extent is very difficult to assess, as is its impact on populations. Hunters rarely keep records, so the evidence linking declines in Osprey numbers to shooting has usually been vague and circumstantial.

Why shoot an Osprey, an enlightened citizen of the twenty-first century might ask in retrospect? To a fisherman or a fish farmer, a person whose livelihood depends on a small pond or hatchery, an Osprey can be a threat. Conflicts with fish pond owners are probably an underrated problem posed by Ospreys and are likely to become more serious as fish farms proliferate, especially in developing countries. Bird hunters have sometimes viewed Ospreys as vermin or just a tempting target. Bounties encouraged this view in European countries during the nineteenth and early twentieth centuries (Bijleveld, 1974). For other people, Latin American peasants for example, Ospreys may be meat for the pot. And for a dwindling few, a dead Osprey may possess magic. A missionary reported how a Bolivian Indian, armed with a bow and arrow, shot an Osprey and then slipped a warm bone from the hawk under the skin of his forearm, apparently in the hope of absorbing hawklike skills at hunting (R. Taylor, unpublished).

If European hunters had resorted to such painful rituals, more Ospreys would remain on the continent today. European raptors were hunted intensively during the eighteenth, nineteenth, and early twentieth centuries, but the efficiency of that persecution increased dramatically after 1850 when accurate breech-loading guns began to replace muzzle-loaders. Bijleveld (1974) has recorded the history of this carnage and it makes depressing reading. Over a period of 20 years in the late nineteenth century, one German keeper shot almost 200 Ospreys at fish ponds under his care, many during migration. During the fall of 1953, 93 migrant Ospreys were shot at a few ponds in lower Saxony. Pole traps took their toll as well; 20 Ospreys were caught in a single year at traps alongside a pond in German Mecklenburg. Unrecorded killings on a similar scale were probably frequent. Such figures emphasize that deaths by shooting in a single locale can be numerous, especially where ponds continually attract new victims.

While there is little evidence that killing migrants reduced Osprey breeding numbers, shooting at nesting sites drastically cut populations in some cases. During World War II, when Europeans were apparently too preoccupied to kill Ospreys or rob their nests, the much persecuted breeding colony on the Darsz Peninsula of northern Germany grew from about six to 22 pairs. After the war, hunting increased again and the colony disappeared by the 1960s (Bijleveld, 1974). In Scandinavia, protection for locally depleted populations in Finland and Sweden arrived during the 1920s and 1930s, and these birds quickly regained their numbers (section 3.1). In Norway and Denmark, by contrast, protection did not begin until the 1950s and 1960s and was often ineffective even then (Bijleveld, 1974). These Osprey populations have recovered more slowly.

Elsewhere in Europe, the regions where Ospreys disappeared as breeders or were most reduced are those where shooting of nesting birds was most intense: the Mediterranean countries, much of central Europe, and Britain (Bijleveld, 1974). In central Europe, only the East German population has remained numerically strong, perhaps because many of these birds nested close to villages and were considered pets (Moll, 1962).

North Americans, it seems, never shot Ospreys with the fervor Europeans did. In the northeastern United States, some local Osprey populations were threatened by nineteenth and early twentieth century sportsmen, but most remained relatively stable before the introduction of pesticides around 1950 (section 9.6; Chapter 11). In northwestern Baja California, shooting probably helped to eliminate most breeders by the 1940s (section 3.4). Elsewhere in North America, Ospreys were generally tolerated – shot occasionally, perhaps, but seldom systematically. Reports of shooting in other parts of the world have also been rare, although the USSR, the Caribbean, and the Middle East may be exceptions. So much of this latter information is sketchy and anecdotal, however, that no accurate estimates of shootings can be made.

As a general rule, people have shot many more Ospreys on migration than on the breeding grounds, probably in part because the birds migrate during fall hunting season, often traveling through countries where raptors are 'fair game'. Young Ospreys migrating for the first time are the most vulnerable segment of a population and have formed at least 80%–90% of the hunters' bag (Melotti & Spagnesi, 1979). Interestingly, such killing never seems to have had much impact on breeding populations. During a period when at least 6% of the Swedish young were shot on migration each year (Table 9.1), for example, Swedish breeding populations were relatively stable (Odsjö, 1982). This suggests that, away from the breeding grounds, hunters take mostly 'surplus' birds. Shooting rates must be very high, apparently, before the populations supplying those migrants are affected. Shooting on the nesting grounds, by contrast, strikes at the breeding adults, the most productive and least expendable segment of a population. In considering the overall impact of shooting on an Osprey population, therefore, it is important to note which birds are killed, not just how many.

Hunting pressure on Ospreys in both Europe and North America has relaxed in recent decades. Two to three times fewer Swedish migrants were recorded shot or trapped in Europe after 1970 than

Table 9.1. *Changes in the percentage of Ospreys from Sweden and the United States shot or trapped before and after 1970. European band recoveries were of Swedish and Finnish young taken in Europe during their first fall migration. Other band recoveries were of US Ospreys of all ages divided into two groups: those taken in the USA and those taken in Latin America.*

	Percentage shot or trapped	
Region	before 1970	1970–1984
Europe[a]:		
of all birds recovered	70	20
of all birds banded	5–6	0.5–1.5
USA[b]:		
of all birds recovered	29	2
of all birds banded	2.6	1.8
Latin America[b]:		
of all birds recovered	18	28
of all birds banded	1	1.5

[a] Before 1970 = 1945–1970.
[b] Before 1970 = 1930–1970.
Odsjö (1982), Saurola (1975), Poole & Agler (1987).

before (Table 9.1). Many of these birds were killed (along with other raptors) in Italy, but this slaughter dwindled after the Italian government passed protective legislation during the early 1970s (Melotti & Spagnesi, 1979). Thanks to education campaigns aimed at hunters, recoveries of Ospreys shot in the United States also declined during the 1970s, although reported shooting rates in Latin America did not (Table 9.1). Raptors are rarely protected in Latin America. Elsewhere, Ospreys are now protected in all countries where they breed except Greece and Iran (Conder, 1977; Hilton, 1977), but local laws often go unenforced.

9.3 Nest robbing

De Naurois (1969) visited the Cape Verde Islands during the 1960s and described villagers there making annual trips to nearby Osprey nests to collect eggs and young for food. He wrote that the islanders considered young Ospreys to be 'les mets de choix' (choice morsels). Arab Bedouins are also thought to rob nests for the pot (Zimmerman,

1984). Osprey eggs and young are thus readily eaten in certain areas, even though they have, to my nose, a decidedly rank odor about them.

It is difficult to know how nest robbing alone has affected Osprey populations because people robbing nests often shoot the adults as well. There is little mention of Cape Verdeans shooting Ospreys, however, so there the species has apparently survived decades of sporadic nest robbing, probably because so many nests are inaccessible. Indeed, nest robbing has undoubtedly helped to select for the high cliff nests that now predominate on these Atlantic islands and along nearby Mediterranean shores. Only in Britain have Osprey nests been robbed often enough to reduce breeding numbers (Chapter 11).

9.4 Human disturbance

In many parts of the world, as we have seen, the Osprey is now a dooryard bird, a bird of the suburbs, regularly feeding and nesting alongside humans (Fig. 9.1). At the same time, Ospreys continue to breed in undisturbed wilderness habitat. Because of this varied exposure to people, and because many other raptors are known to suffer from activity near their nests, people have wondered how much disturbance nesting Ospreys can tolerate.

Researchers survey Osprey nests in one of two potentially disturbing ways: from the ground or from a low-flying airplane. Ground surveys involve climbing to nests via a ladder (Fig. 9.2), mechanical means (Fig. 9.3), or the nest tree itself. A pole with a mirror is sometimes used when a nest is low but not otherwise accessible (Fig. 9.4). Nest visits give accurate counts of eggs and young and also allow collection of prey remains, addled eggs, or data on the growth and condition of young. Ground surveys take longer than aerial surveys, however, and are thus potentially more disturbing to the birds.

To determine the impact of ground surveys on breeding Ospreys, I compared reproductive success at nests subjected to varying levels of disturbance by my own and other research (Poole, 1981). Nest visits occurred throughout the nesting season and generally lasted 20–30 minutes. Breeding rates were just as high at nests visited two to five times as at those unvisited and undisturbed, where young were counted from a distance through a telescope. I was careful not to visit nests during cold, rainy, or very hot weather, however, especially when small young were in the nest. Heat is particularly stressful to eggs and young (Drent, 1975). In addition, I avoided climbing trees

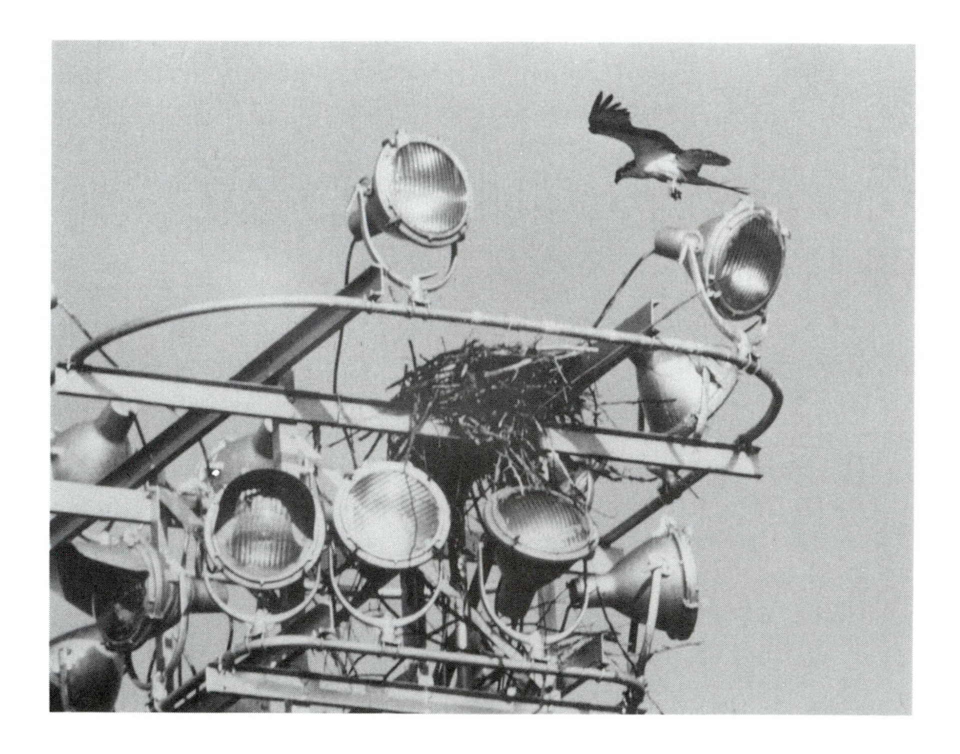

Figure 9.1. Two views of an Osprey nest on a light tower at a busy urban parking lot along the Connecticut (USA) coast. This nest has regularly fledged young during the past decade. (Photos: P. Spitzer (bottom); B. McCormick (top).)

in areas harboring Raccoons because these predators can use human scent trails to locate bird nests. I also avoided visiting nests when young were near fledging because it is easy to flush such young prematurely, before they can fly well enough to return to the nest. Using proper precautions, quick visits to Osprey nests apparently do not influence reproductive success.

Figure 9.2. Checking the contents of an Osprey nest on a New England salt marsh. If proper precautions are taken, nest visits like this have no harmful effect on Osprey breeding success. (Photo: D. Haney.)

Aerial surveys require slow flights over nests with a fixed-wing aircraft or brief hovers with a helicopter. Where nests are widely scattered or inaccessible from the ground, aerial surveys are the only efficient way to find nests and to check on activity and success, even though eggs and young are often difficult to count from the air and

Figure 9.3. Checking Osprey nests near live electrical wires can present problems. With the cooperation of a local electrical company, however, a boost from a hydraulic bucket truck overcomes the difficulties. (Photo: A. Poole.)

flight time is expensive (Carrier & Melquist, 1976). Helicopter surveys of more than 200 Osprey nests in northern Idaho (USA), showed that adults returned to nests quickly (between one and three minutes) when flushed, that levels of reproductive success and nest abandonment were comparable to other populations not surveyed by

Figure 9.4. Attaching a mirror to a long pole provides a quick and simple way to check the contents of Osprey nests that are not too high. (Photo: A. Poole.)

air, and that Ospreys successfully avoided the aircraft despite occasional dives at it (Carrier & Melquist, 1976). There is one report of an Osprey killed by a survey plane that the bird attacked (Scott & Surkan, 1976), but such collisions are apparently rare. Although aerial surveys are noisy and potentially disturbing, therefore, they have little effect on Osprey reproductive success. In fact, one of the main problems with aerial surveys is that incubating adults often refuse to leave the nest despite the clatter of a helicopter hovering nearby, frustrating a count of eggs or young.

Human activity near Osprey nests might seem less directly disturbing than nest visits or aerial surveys, but it is potentially more disruptive because people often linger near nests, unaware of the disturbance they are creating. Some reports have shown that Ospreys nesting near people reproduce poorly; other have found no negative effects. The key to these differences seems to lie in the timing of disturbance and the extent to which Ospreys are habituated to it. Among Ospreys in the eastern United States, for example, I found that pairs breeding in suburban habitat (less than 300 meters from roads, railroads, boat channels, or inhabited houses) raised as many young as pairs breeding within large natural reserves rarely visited by people (Poole, 1981). Significantly, these suburban Ospreys were exposed to relatively continuous human activity from the time they arrived and started building nests. Good examples of such tolerance are the several pairs of Ospreys in southern New England that have built their nests on telephone poles along the main line of the railroad between New York City and Boston. As the trains rumble past, the incubating birds usually sit tight, clearly habituated to the sporadic appearance of these one-eyed, steel giants.

Not all Ospreys are tame, however. Swenson (1979a) found that pairs breeding near campsites in Yellowstone National Park (Wyoming, USA) reproduced poorly compared with pairs breeding in more isolated sections of the park. These campsite pairs were exposed to abrupt increases in disturbance when camping and fishing seasons opened each year, about midway through incubation, so the sudden appearance of people near nests was plausibly the critical factor here. Likewise, Ospreys in northern California bred least successfully near logging roads; logging usually began well after eggs were laid (Levenson & Koplin, 1984). In Scandinavia, an increase in boating each spring was thought to threaten the success of nests on certain lakes (Haga, 1981; Hallberg, Hallberg & Sondell, 1983). Habituation is the key to Osprey tameness, therefore, and nesting pairs disturbed only sporadically are those most in danger of failing and thus most in need of protection.

None of the studies mentioned noted how long adults were kept off the nest by humans, but the assumption has been that eggs or young at disturbed nests are not properly incubated or brooded, and so die of exposure. Alt (1980) watched nesting Ospreys react to boats along Wyoming rivers. He found that boats traveling parallel to the shoreline did not flush incubating birds while boats equally far away, but heading toward nests, did. Ospreys remain alert to subtle differences in human behavior, therefore, and distance from nests is not the sole criterion to be considered in establishing management guidelines for this species.

The Osprey's potential tameness has had a major impact on its distribution, allowing it to breed successfully in settled areas where many other large birds of prey would never venture. Such tameness, in fact, has helped to make the Osprey the ecological opportunist that it is. Gerrard *et al.* (1976) showed that Ospreys outnumber Bald Eagles in areas of central Saskatchewan where roads give people access to lakes. At more remote lakes, however, the eagles outcompete Ospreys for food and nest sites. Tolerance of people, therefore, will be a key to the Osprey's continued success as a species.

9.5 Forest clearing and habitat degradation

In many regions where Ospreys nest, especially Europe and the northwestern United States, forests have been heavily managed for timber. As a result, dead snags, tall trees, and forests bordering lakes and rivers have often been cut, destroying existing and potential nest sites and silting up the waters where Ospreys fish. Centuries of forest clearing in central and eastern Europe have made nest trees especially scarce, probably one reason that many European Ospreys now nest on artificial sites. Foresters have also cleared or thinned large areas in Fennoscandia and the northwestern United States, but much woodland remains and the negative effects of such harvesting have been compensated to some extent by artificial nesting platforms (Chapter 10). Foresters now realize that by leaving nest trees and strips of timber along shorelines, the harmful impacts of timber operations on Ospreys can be mitigated to some extent (Garber, Koplin & Kahl, 1974).

Ecological changes more subtle than timber harvesting have also affected Osprey status and breeding success. Ospreys are abandoning parts of Florida Bay, for example, in part because the natural flow of fresh water to that estuary has been altered by dams and channels, reducing populations of fish (Kushlan & Bass, 1983). Farther north, Chesapeake Bay has recently experienced severe changes in its ecology due to excess sewage and fertilizers – nutrient overloading; loss of bottom vegetation and fish kills have been symptoms of this stress (Officer *et al.*, 1984). Some Chesapeake Ospreys have had difficulty finding fish, and their nestlings have starved as a result (section 7.2).

Perhaps more worrisome is the widespread clearing of tropical forests where northern Ospreys winter and some Asian Ospreys breed. The world's tropical forests, now being cut at the alarming rate of about 2% annually (about 5% annually in West Africa and Southeast Asia), will disappear by the year 2000 if current rates of clearing continue (Brown *et al.*, 1986). Clearing triggers erosion and runoff, often clogging rivers with silt, and damaging reefs and coastal waters that sustain key fisheries. As a result, Ospreys become more crowded, more dependent on fish ponds, and thus more vulnerable to shooting. Similar results are likely as large tracts of coastal mangrove forests in Southeast Asia and Latin America are converted to fish and shrimp farms (Rabanal, 1978).

9.6 Organochlorine pollutants

Because birds of prey are highly visible animals that react strongly to many pollutants, they have provided clear, well-studied examples of the ways in which toxic substances accumulate in ecosystems and cause harm. By using the Osprey's reproductive success as a measure of environmental health, and by comparing this with similar measures provided by other birds, it has been possible to answer several key questions about the ecological effects of persistent toxic chemicals:

(1) What regions of the northern hemisphere have been most contaminated, and by what pollutants?;
(2) What harm have different pollutants brought to breeding Ospreys and other birds?;
(3) Have migratory birds suffered most from pollutants accumulated on their breeding or wintering grounds?

Although Ospreys gained fame as pollution victims, it is surprising how few populations were actually affected by toxic chemicals. Those populations that did suffer from pollutants were primarily

North American ones, surrounded by people interested in their fate who publicized it widely. From a global perspective, however, most Ospreys continued to produce young at normal rates throughout the worst of the pesticide years.

Uses and dangers

Of all toxins, the organochlorine compounds have harmed Ospreys the most. The simple laboratory procedure of attaching chlorine to carbon unleashed greater ecological damage, over a shorter period of time, than decades of sporadic shooting, nest robbing, and habitat destruction had done. The most potentially stressful of these compounds for Ospreys, in part because the birds accumulated them in the greatest amounts, included DDT (dichloro-diphenyl-trichloroethane), the cyclodienes such as dieldrin, aldrin and heptachlor, and various of the polychlorinated biphenyls (PCBs). DDT and the cyclodienes were used primarily as insecticides; PCBs were industrial compounds used mostly as fluids in electrical equipment. Use of these chemicals accelerated after World War II, but no one fully realized what ecological damage they were causing until the 1960s.

Several characteristics common to the organochlorines make them all potentially harmful to a top predator like the Osprey (Newton, 1979). First is their stability. They can take decades to break down fully in the environment. Second is their tendency to disperse widely, whether attached to sediment particles, swept along by water currents, or volatized and carried aloft by winds. Third, and particularly important for raptors, is the lipophilic nature of these substances, that is, their tendency to become trapped in fatty tissues and thus to concentrate biologically in successive levels of a food web. This means that predatory and fish-eating birds often accumulated levels of organochlorines up to a million times greater than background levels in water and up to a hundred times greater than levels in the food they ate (Woodwell, 1967; Newton, 1979). Fish concentrate such pollutants in their fat because they absorb them directly from water through their gills, supplementing what they get from contaminated food. A fourth danger of some organochlorines is their ability to disrupt breeding by reducing egg viability at very low concentrations, as low as a few parts per million (ppm) wet weight in fresh eggs. Raptors seem more sensitive than most other birds to such effects. Lastly, because manufactured organochlorines are recent artificial substances that have played no part in long-term evolution, birds are seldom able to metabolize or excrete them efficiently. Thus, the higher the level in a bird's diet, the higher the level that is generally found in its tissues and eggs. Females do excrete a portion of their body's pesticide burden in the yolk of their eggs (Bogan & Newton, 1977), but while this may temporarily cleanse the female herself, it jeopardizes the success of her clutch.

By the early 1970s, various human health and environmental considerations had encouraged the nations of Europe and North America to restrict or ban the use of the most toxic, persistent organochlorines. Insect resistance, contamination of human food, and adverse publicity (especially in the wake of Rachel Carson's book *Silent Spring*) had overcome years of propaganda by the chemical industry. Concern over the impacts of organochlorines on nontarget organisms like Ospreys played a noisy part in this debate, actually tipping the scales toward restricted application in a few regions of the northeastern United States (Chapter 11), but risks to human health were the impetus behind most restrictions. Manufacturers turned increasingly to the production of pesticides that had specific targets and that broke down quickly in the environment. Nevertheless, many of the broad spectrum, long lasting pesticides like DDT and dieldrin are still used widely in developing countries, particularly in tropical regions where Ospreys winter. Even in areas where organochlorines have been restricted, economically important uses are occasionally allowed, stockpiled supplies are used illegally, or the substances are found as trace contaminants in legal pesticides. Many Ospreys, therefore, continue to receive some exposure to organochlorine pollutants that were banned decades ago.

Impacts on reproduction

Evidence that Ospreys were accumulating organochlorines first emerged during the mid-1960s in the northeastern United States. Peter Ames (1966) collected a series of eggs from nests in coastal New England where few eggs were hatching and the Osprey population was dwindling rapidly. For comparison, he also collected eggs from a relatively healthy and stable population in Maryland, several hundred kilometers farther south. When analyzed, all eggs contained measurable levels of DDT and its main metabolite, DDE. Eggs with the highest levels came from states with the lowest reproductive success. Maryland eggs, for example, were about 40% less contaminated than Connecticut eggs and produced about twice as many young. In addition, fish eaten by Connecticut Ospreys contained five to ten times more DDE than Maryland fish did, reflecting differences in local levels of pesticide use and clearly suggesting that Ospreys had been contaminated by their diet.

What this early study failed to recognize, however, was the crucial link between DDE and reproductive failure – eggshell thinning. It remained for Ratcliffe's (1967) pioneering studies of British raptors to show convincingly that eggshell thickness declines with rising levels of DDE in eggs. Alerted to this problem, people noted sudden and unprecedented declines in shell thickness among European and North American Osprey populations during the 1960s (Figs. 9.5 and 9.6; Wiemeyer *et al.*, 1975; Spitzer *et al.*, 1977). Nests often contained cracked eggs and shell fragments during those years. Nearly all populations examined were affected to some degree. Average declines in shell thickness for different populations ranged from 1% to 18% below pre-DDT norms (Appendix 7). Regions exposed to heavy pesticide use were generally those showing the most extensive shell thinning. Only regions remote from DDT

Figure 9.5 In study nests, eggs that fail to hatch are usually those collected for analysis of contamination. (Photo: A. Poole.)

applications, Baja California and south Florida for example, had negligible contamination and thinning. Importantly, whole populations were affected, not just a few individuals (Fig. 9.6). Other predatory and fish-eating birds also accumulated DDT residues and showed some of the same trends (Newton, 1979; Nisbet, 1980).

As data accumulated, the relationship between shell thinning and DDE in Osprey eggs became clearer. Although shell thickness declined significantly with increasing amounts of DDE in the egg, the relationship between these two parameters was not linear. Instead, as typically seen in other birds sensitive to DDE (Blus *et al.*, 1972; Newton, 1979), a roughly linear relationship could be shown between shell thickness and the logarithm of DDE levels in Osprey eggs (Fig. 9.7). This meant that changes in thickness at lower levels of contamination were much greater than those at higher levels. As DDE levels in eggs increased from 0 to 12 ppm, for example, shell thickness dropped about 14% (on average), while all further contamination brought only a 6% drop in thickness (Fig. 9.7). Such a pattern indicates that shell thickness, and thus egg viability, can change rapidly with only small fluctuations in DDE contamination. This was one reason why Osprey population crashes during the

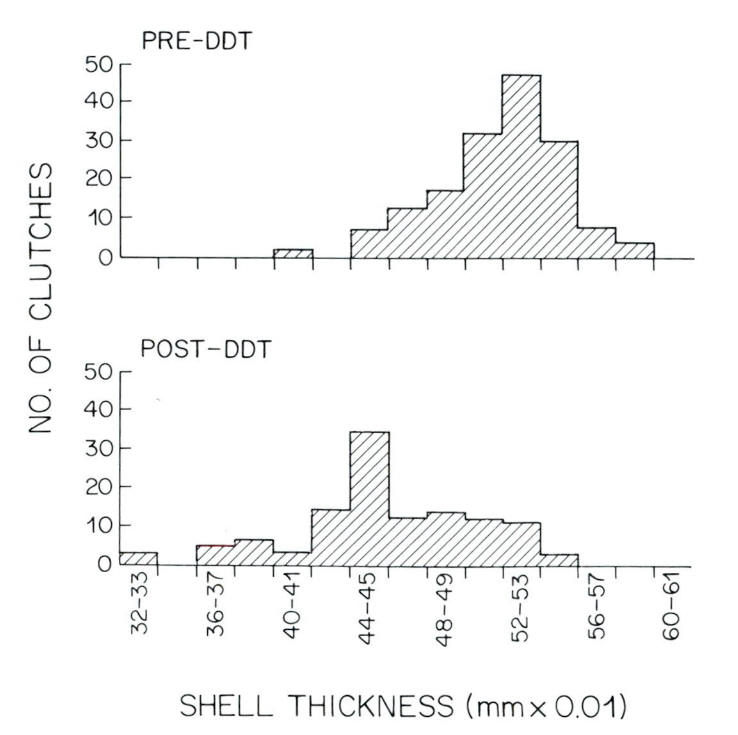

Figure 9.6. Historical changes in shell thickness of Swedish Osprey eggs, before and after the introduction of DDT in 1946. Pre-DDT thicknesses are based on measurements from museum egg collections, 1861–1946. Post-DDT measurements are from unhatched, intact eggs collected from 1962 to 1973. DDE (the main metabolite of DDT) thins shells by inhibiting enzymatic (ATPase) activity in the shell gland, which prevents transport of calcium across membranes to the shell (Peakall, 1975). Data from Odsjö (1982).

Table 9.2. *Thickness of eggshell fragments found in Swedish Osprey nests (1971–1973) in relation to the number of eggs per nest remaining intact throughout incubation. Shell thickness varies little among eggs in a clutch, so even a few shell fragments provide a reliable measure of shell thickness for an entire clutch. Three eggs was the most common clutch size in this Swedish population.*

No. of intact eggs/nest	No. of clutches	Mean shell thickness (mm)	% decrease in thickness compared with pre-DDT eggs
0	5	0.366	29
1	14	0.395	23
2	34	0.431	16
3	72	0.452	12

Data from Odsjö (1982).

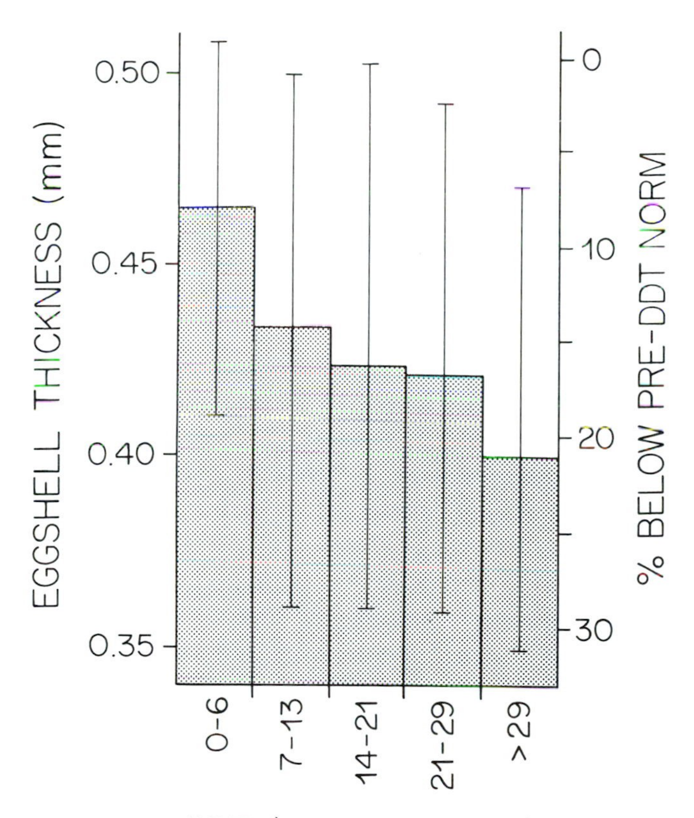

Figure 9.7. Osprey eggshell thickness in relation to DDE residues in the egg. Data shown are mean values (vertical lines show ranges) based on analyses of 112 eggs, most collected in the northeastern United States during the 1960s and 1970s. Data from Spitzer *et al.* (1978) and A. Poole & J. Farrington (unpublished).

1950s and 1960s, and recoveries during the 1970s, were so sudden. Small differences in contamination were often enough to spell success or failure for a population.

Individual variation in Osprey eggshell thinning at any one level of DDE contamination can be considerable, as seen in the range of values shown for each histogram in Figure 9.7. There is no guarantee, therefore, that one particular concentration will produce the same level of shell thinning in each female. Some females may simply be less susceptible to the effects of DDE than others. In addition, shell thickness varies naturally among individuals in uncontaminated populations (Fig. 9.6, top).

Thin shells increase the chances that an egg will break before hatching. Odsjö (1982) measured the thickness of shell fragments from Swedish nests containing different numbers of eggs that remained intact through incubation. He found that nests where most eggs broke contained thinner fragments than those where two or more eggs hatched (Table 9.2). Such data help to show where the critical levels of shell thinning lie for Ospreys. Eggs 10%–12% thinner than pre-DDT norms seldom break. Clutches with eggs thinned more than about 16%, however, are likely to lose at least one egg, and egg loss accelerates rapidly with greater thinning.

Because Osprey eggs break regularly when shells are about 16% thinner than normal, one might expect that only those levels of DDE capable of producing such thinning (about 15 ppm wet weight; Fig. 9.7) would seriously affect reproduction. As Figure 9.8 makes clear, however, once DDE in eggs exceeds about 5–10 ppm, even successful pairs rarely produce more than one young per year, so a population's overall breeding rate is too low to balance adult mortality. Since eggs with less than 15 ppm DDE seldom break, loss of breeding potential is mostly due to hatching failure. Infertility or egg neglect can prevent eggs from hatching, but these factors were no more prevalent among contaminated Ospreys than among normal birds (Wiemeyer *et al.*, 1975). Instead, people often found dead, partially developed embryos in unhatched Osprey eggs, suggesting that DDE or other contaminants had poisoned the eggs.

The problem in singling out one contaminant for blame was that levels of different organochlorines in Osprey eggs were often correlated, making it difficult to separate toxic effects. In addition, combinations of organochlorines can be more toxic than any one compound occurring alone at similar levels (Risebrough & Anderson, 1975). Finally, DDE disrupts the pore structure of the eggshell, reducing gas exchange between a growing embryo and the atmosphere and thus jeopardizing embryo viability even if the egg never breaks and other contaminants are absent. Although egg breakage was an important cause of Osprey breeding failure during the DDT era, it was certainly not the only cause.

Dieldrin is very toxic to brain and other nervous tissue. Many birds – eagles that ate contaminated sheep carcasses and seabirds that ate contaminated fish – have died from dieldrin poisoning (Newton, 1979; Nisbet, 1980). Despite this potential toxicity, few Ospreys are known to have suffered from dieldrin. The highest levels of dieldrin in Osprey eggs were found in Connecticut and coastal Virginia during the late 1960s (up to 2.8 ppm; Wiemeyer *et al.*, 1975; Henny *et al.*, 1977; Spitzer *et al.*, 1978). Both populations were reproducing poorly at the time, but their eggs also contained DDE, so the effects of the two compounds could not be separated. Eighty-eight percent of 26 dead Ospreys examined during the 1960s and 1970s in the eastern United States carried detectable levels of dieldrin in their tissues, a few with brain dosages high enough to have contributed to their deaths (Wiemeyer, Lamont & Locke, 1980). Overall, however, dieldrin appears to have been a more localized and much less critical contaminant for Ospreys than DDT.

Even though PCBs are thought to have caused hatching failure and embryonic defects in several species of birds (Hays & Risebrough,

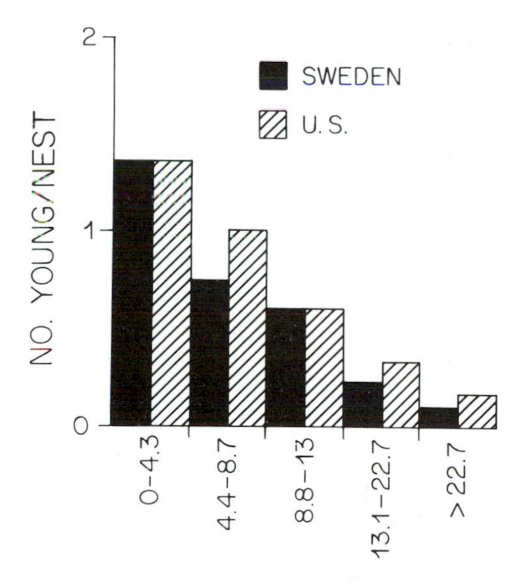

Figure 9.8. Mean brood size at Osprey nests in Sweden (1971–1973) and in New England (USA) (1969–1984), in relation to DDE levels in unhatched eggs from those nests. Swedish data, from Odsjö (1982), show young hatched per nest ($N = 108$ nests); US data, from Spitzer *et al.* (1978) and Poole & Farrington (unpublished), show young fledged per nest ($N = 101$ nests).

1972; Risebrough & Anderson, 1975), Ospreys can apparently tolerate high doses with little effect on breeding rates. The Osprey colony I study in southeastern Massachusetts, for example, has been heavily contaminated with PCBs for the past decade, but not other organochlorines. When last sampled (1980–1983), PCBs in addled eggs of these birds averaged 25 ppm wet weight, a high level for any wild bird. Yet at the same time these Ospreys have produced young at rates higher than any other Osprey population in the world (Appendix 4), and adult mortality has remained at normal levels for the species. Furthermore, Osprey populations in Connecticut and Long Island recovered rapidly after DDT was banned, despite continued high levels of PCBs in their eggs (Spitzer *et al.*, 1978). And in Sweden, PCBs were equally abundant in successful and unsuccessful clutches, as long as clutches with complicating levels of DDT were removed from the sample (Fig. 9.9, bottom). Isolated from other contaminants, therefore, it seems that even high levels of PCBs fail to harm Osprey eggs.

Different species, apparently, have very different tolerances of PCBs. Chicken eggs containing only a few ppm PCBs rarely hatch (Scott *et al.*, 1975) and are banned from sale in Europe and the United States because of human health considerations. Ospreys, apparently, are among the few lucky species unaffected by PCBs.

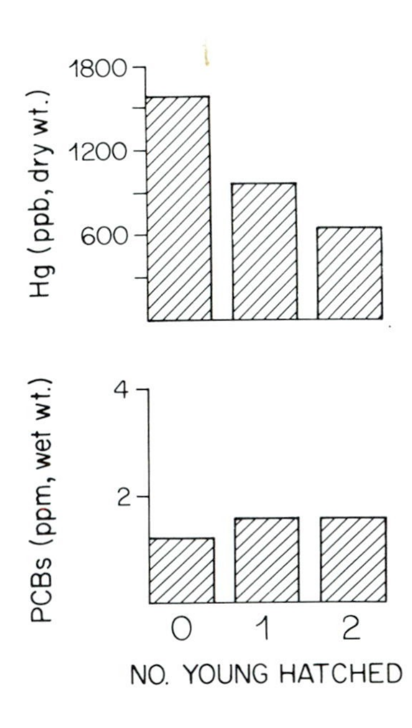

Figure 9.9. Mean levels of mercury (Hg) (top) and PCBs (bottom) in eggs of Swedish Ospreys during the 1970s, in relation to the number of young hatched at nests where eggs were sampled. Eggs with high levels of DDT (more than 2.3 ppm, wet weight) were removed from these samples in order to study the effects of PCBs and mercury in isolation. Note that PCBs were uniformly low for all brood sizes, suggesting little effect on egg viability, while mercury levels were high at nests that failed. From Odsjö (1982).

Population effects

Clearly, a population cannot sustain its numbers long without adequate reproduction. While Osprey populations do weather occasional poor seasons with negligible long-term effects on numbers, pesticides were relentless. Season after season of poor breeding success meant that few young Ospreys were available to replace breeders lost to yearly mortality. As a result, the age structure of heavily contaminated populations was skewed increasingly toward older birds, and breeding numbers dwindled. Immigration never compensated for such losses because Ospreys in these localities seldom dispersed far from breeding or natal sites (Fig. 8.1).

The hardest-hit Osprey populations were those where DDT use and consequent shell thinning were greatest. Of those populations known (or calculated from reproductive rates) to be declining during the pesticide era, most produced eggs with shells thinned to critical levels (Fig. 9.10). The greatest losses occurred along the northeast coast of the United States, where reproduction was sufficiently low over a period to reduce breeding numbers 50%–90% in various localities (Chapters 3 and 11). The salt marsh ecosystems supporting these Ospreys were sprayed directly with DDT, so it is not surprising the birds fared so poorly.

Elsewhere in the United States, few other populations were affected as drastically. Ospreys breeding along Chesapeake Bay and in northern Michigan did fail repeatedly during the 1960s, but these populations apparently lost fewer breeders than those in the north-

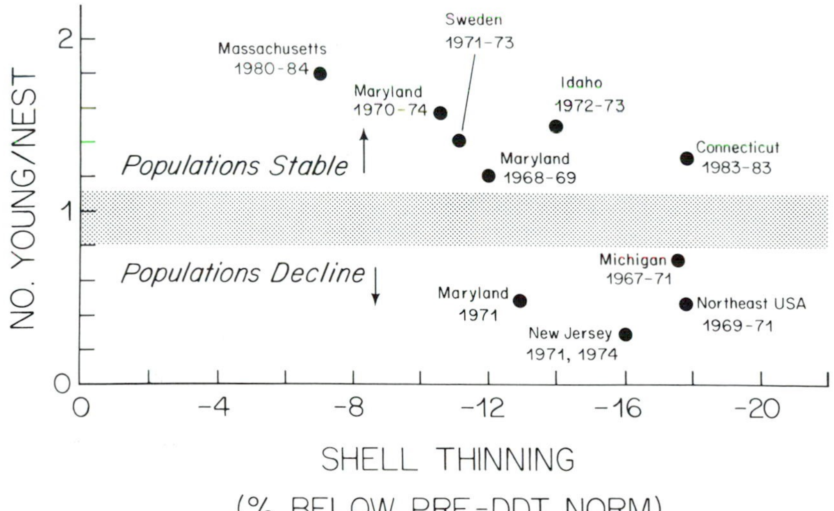

Figure 9.10. Osprey reproductive success and population status in relation to eggshell thinning. Dotted band shows the number of fledged young thought necessary to balance adult mortality in different Osprey populations (see Chapter 8). References in Appendix 7, except Maryland 1970–74 from Reese (1977); Michigan 1967–71 from Postupalsky (1977); Maryland 1971 from Wiemeyer (1977).

east (Chapter 3). Far fewer Ospreys were analyzed for pesticides in Europe, in part because most European populations remained stable during the pesticides era, so there was less cause for concern. At lakes in southern Sweden, reduced hatching rates were associated with moderate levels of DDT in eggs (Appendix 7), but overall this population suffered few losses during the 1950s and 1960s and continued to produce young at rates close to historical levels (Österlöf, 1973; Odsjö & Sondell, 1976). It seems likely that Swedish Ospreys breeding along the Baltic coast were more heavily contaminated, given the very high levels of organochlorines found in White-tailed Sea Eagles from that region (Jensen *et al.*, 1972), although Ospreys from the Gulf of Bothnia were almost an order of magnitude less contaminated than neighboring sea eagles (Koivusaari *et al.*, 1972). Sea eagles eat carrion and thus are more susceptible to accumulating toxins.

Britain's Ospreys, like Sweden's, showed no consistent reproductive failure due to pesticides, perhaps because most nested in areas where few pesticides were used (Dennis, 1983). The most polluted Ospreys in Europe may have been those breeding along Mediterranean shores. Very high levels of DDE were found in two Osprey eggs from Corsica, the only Mediterranean eggs analyzed to date (Appendix 7). Osprey numbers have declined drastically in Mediterranean countries during recent decades, but factors other than pesticides are also to blame (Chapter 3). In short, it seems fair to say that most European Ospreys escaped the trauma of pesticides, probably because their main feeding habitats were seldom sprayed or treated directly as in the United States.

All of the evidence implicating pesticides in Osprey reproductive failures was circumstantial. No one brought Ospreys into the laboratory, fed them a diet with known levels of contamination, and measured the results. This was done with other raptors, however, confirming what field studies had suggested (Newton, 1979). Nevertheless, as many uses of organochlorines were phased out during the 1960s and 1970s, these changes functioned as large scale experiments, albeit poorly controlled ones. Background environmental levels of most organochlorines dropped quickly, probably in part because sunlight and soil microbes were more efficient degraders of these compounds than people had suspected (Crosby, 1973; Fleming, Clark & Henny, 1983). Ospreys showed correspondingly quick recoveries. By the late 1970s, nearly all previously contaminated North American Osprey populations were reproducing at close to pre-pesticide rates and fast regaining breeders (Henny, 1977a & 1983; Spitzer *et al.*, 1978).

9.7 Mercury

Mercury has been a problem for Ospreys in some regions because, like the organochlorine pesticides, it is soluble in fat and blood, it concentrates rapidly in aquatic food webs (often absorbed directly from water to blood through gill membranes), and it is toxic at very low levels (NRC, 1978). Unlike the organochlorines, however, mercury can be excreted with some efficiency. Birds do this by dumping substantial quantities of mercury from blood to growing feathers (Berg *et al.*, 1966), thus removing the toxin to an inert portion of the body where it can eventually be molted. When feathers are not growing, however, or when levels in a bird's diet are particularly high, mercury does accumulate in body tissues or eggs, potentially harming both.

Only in Scandinavia and the United States have people looked for mercury in Ospreys, mostly during the 1960s, when discharge of waste mercury by industry was at its peak, and during the 1970s, after discharge was banned (Appendix 8). Besides analyzing eggs, chemists checked the mercury content of feathers from museum skins – birds collected at known dates. This provided an ingenious way of tracing historical changes in the levels of mercury in the environment. From 1840 to 1940, for example, mercury in the feathers of Swedish Ospreys was barely above natural background levels. From 1940 to 1965, feathers were two to three times more contaminated, on average, reflecting increased discharge of mercury by Swedish pulp and paper industries (Jensen *et al.*, 1972). When Finland banned waste discharge of mercury, levels in Finnish Osprey feathers declined significantly (Häkkinen & Häsanen, 1980).

In general, Swedish eggs have been more contaminated than eggs from Finland or the United States (Appendix 8). Hatching rates were depressed during the early 1970s at a few locations in Sweden where mercury levels were especially high (Fig. 9.9, top). A few Ospreys found dead in the eastern United States may have died from mercury accumulation (Wiemeyer *et al.*, 1980 & unpublished). Significantly, all studies found mercury contamination to be a localized phenomenon. Usually Ospreys nesting on just a few lakes or rivers accumulated high doses. Considering how rapidly aquatic animals absorb mercury, this is perhaps not surprising. In its soluble (and most toxic) form, mercury seldom travels far from a dispersal point before being bound up in blood (NRC, 1978).

9.8 Pollutants on the wintering grounds

Although the breeding grounds of European and North American Ospreys have become less contaminated since the 1970s, organochlorines and heavy metals are still used extensively in some tropical and subtropical countries where northern populations winter and resident Ospreys breed (Linear, 1982; Mowbray, 1986). During the height of the pesticide years, it was assumed that migrant Ospreys accumulated most pesticides at breeding sites because, despite sharing common wintering grounds and migratory routes, some populations (and their prey) were much more contaminated than others. More recently, as pesticide use in the tropics accelerated, people worried that northern Ospreys might accumulate contaminants during migration or at wintering sites, as some Peregrine Falcons have (Enderson *et al.*, 1982; Springer *et al.*, 1984).

Of the many migratory Osprey populations monitored from 1975 to 1985, however, none are clearly threatened by organochlorine contamination. The wide range of DDE concentrations in eggs from the western United States (Flathead Lake, Montana and Eagle Lake, California) suggest that a few Ospreys in those breeding populations could be exposed to high levels of DDT while wintering in Central America, but most of the birds are still breeding successfully (MacCarter & MacCarter, 1979; Littrell, 1986). If Ospreys are accumulating pesticides during migration, therefore, the levels are apparently low enough, and the effects subtle enough, that most individuals are not threatened – at least in those populations that have been studied.

One reason Ospreys may escape serious contamination at wintering sites is that organochlorines often volatilize quickly in warm tropical climates, reducing runoff to local lakes and rivers (Perfect, 1980). In addition, microbes generally remain active year-round in warm tropical soils and waters, potentially degrading contaminants more quickly than in seasonal temperate climates. Alternatively, pesticide use may be sufficiently localized in the tropics that most Ospreys simply never encounter contamination. Those few migrants whose breeding attempts fail because eggs are poisoned or thinned by tropical pesticides would be overlooked by researchers tending to concentrate on population averages.

Resident Ospreys may suffer from pesticide use more than migrants because residents could be exposed to organochlorines year-round and while breeding. Yet we know little about the status and health of these populations. Studies of populations nesting in the southwest Pacific, the Caribbean, and the Middle East are badly needed.

9.9 Acid waters

Scientists and politicians have spent endless hours debating the sources and dangers of acid rain and snow, but the long-term ecological effects of this phenomenon are only beginning to be understood (Mason & Seip, 1985). Many Ospreys nest in regions where lakes are now heavily acidified: Scandinavia, eastern Canada, the northeastern United States, and Scotland. Fish, sensitive to acidity, are often scarce or missing in such waters (Almer *et al.*, 1974). Despite the threat of reduced food availability, however, most Ospreys breeding near acid lakes continue to reproduce quite well, probably because the negative effects of acidity are often localized (Clum, 1986; Eriksson, 1986). Lakes formed in softer sedimentary rock are generally buffered against rising acidity (as are coastal waters), while lakes in harder bedrock become acid more easily. Thus Ospreys in acid regions are often left with a patchwork of productive and unproductive lakes, and a region's average pH has never been found to correlate significantly with Osprey foraging success (Clum, 1986; Eriksson, 1986).

Only one Osprey population, in the Adirondack Mountains of northeastern New York State (USA), seems to have suffered from increased acidity. Here, lake pH is correlated with the breeding success, but not the foraging success, of these birds (Fig. 9.11). This suggests that contamination rather than lack of fish limits production of young. Heavy metals, especially aluminum, dissolve easily in acid waters (low pH), contaminating the local biota and potentially lowering hatching rates of eggs (Nyholm, 1981). No one has measured aluminum levels in Osprey eggs, however, nor in the fish Ospreys eat, so the precise causes of breeding failure at Adirondack lakes remain speculative.

9.10 Summary of threats

The major threats to Ospreys have come in Europe from shooting and forest clearing and in the United States from pesticides.

European Ospreys were shot and trapped most intensively from about 1850 to 1970; fewer have been shot there and in the United States since 1970. Migrants, especially recent fledglings, were shot more often than breeders, but shooting of breeders had a much greater impact on the stability of populations. When not persecuted, breeding Ospreys habituate easily to people near their nests. Remote nests, rarely visited, are those most threatened by disturbance. Timber harvesting and land clearing have destroyed nest sites in parts of Europe and the United States, but nest platforms have compensated for this to some degree. Adult Ospreys have few natural predators, although large owls may kill breeders occasionally. Eggs and young are vulnerable to a wider range of predators.

Although their harmful effects were much publicized, organochlorine pollutants threatened only a few Osprey populations (those breeding in the northeastern and midwestern United States) during only two decades (1950–1970). DDT caused more harm than other organochlorines by thinning eggshells and killing embryos, cutting reproduction well below replacement levels and reducing populations 50%–90% in two decades. Other contaminants such as dieldrin and mercury have had more local impacts on breeding; PCBs have had no measurable impact at all. Migrant Ospreys probably accumulated most toxins on their breeding grounds, even

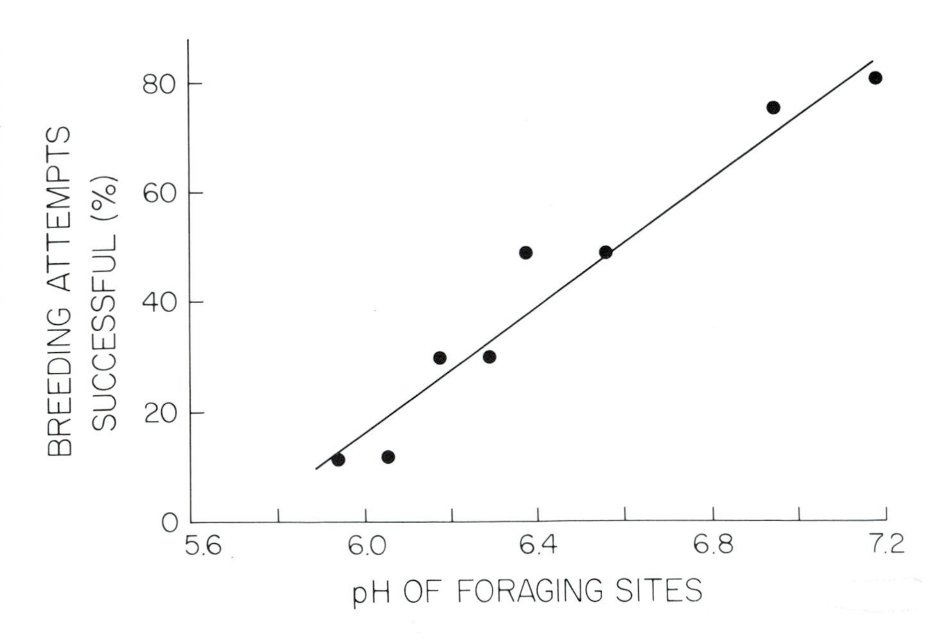

Figure 9.11. Osprey breeding success near Adirondack (USA) lakes with varying levels of acidity. PH levels shown here are averages for lakes within 5 kilometers of nest sites. Each point represents success at one nest site over a period of 3–6 years. From Clum, 1986.

though some organochlorines were (and still are) used heavily in areas where these birds winter. As use of organochlorines dwindled in the United States, Osprey eggs became less contaminated and more likely to hatch. Acid waters, a recent threat to populations breeding along northern lakes, have had only minor effects on Osprey reproduction to date.

10 MANAGEMENT

Increasingly in the future, the survival of species will depend on the extent to which they can make adjustments to radically changed environments and to existence in close proximity to man . . . Some species will be able to make such adjustments on their own . . . Still others will be able to do so with a little intelligent and sympathetic help from men.

Tom Cade (1982)

Even though Ospreys are adjusting quickly to changed environments, nesting comfortably alongside highways, reservoirs, and boat channels, many continue to need intelligent and sympathetic help. Obviously in need are those populations whose numbers have been decimated by man's ignorance or greed. But others, even our backyard pairs, also need help at times. Ospreys can become pests, for example, when they raid a fish hatchery or nest on utility poles, shorting out a town's electrical power. Partly to assuage our guilt for past persecutions, partly to keep certain Ospreys from becoming pests, partly to seek publicity, and partly because stewardship of wild animals seems deeply rooted in our nature, people have rushed to manage Ospreys in recent decades, not always directing efforts toward those most in need. Americans have led the crusade and Ospreys have responded remarkably well.

To thrive, Ospreys, like other birds, need food and nest sites (Chapter 8). In addition, population growth depends on dispersal, an ability to colonize new territory. Managing Ospreys, therefore, has usually meant inventing ways to circumvent these three potential limitations. Dispersal and nests site availability have proven easier to manipulate than food supply. By building artificial nest sites for Ospreys, for example, people in dozens of locations have altered the distribution, density, and breeding success of this species (Tables 8.3 and 8.6). And projects that disperse Ospreys artificially by moving young to regions where populations need bolstering are showing great promise of restoring the species to some of its former haunts. Indeed, the full implications of such reintroductions – an ability to re–establish wild populations of Ospreys almost at will where habitat permits – are profound.

This chapter focuses on reintroductions and artificial nest sites, the two most successful aspects of Osprey management, touching more briefly on ways that fish can be farmed for the mutual benefit of Ospreys and humans. The impacts of artificial nest sites on Osprey breeding success and population dynamics are considered separately in Chapter 8.

10.1 Food

Fish ponds and hatcheries act as natural magnets for hungry Ospreys, creating management problems for advocates of bird and fish alike. Ospreys are often shot in such situations (section 9.2). Tolerant fish–pond owners stand to lose stock. Consider an unguarded hatchery near three active Osprey nests. In three months of steady fishing, Ospreys could easily remove 1800 mature fish (Appendix 3). Unfortunately, there is no simple way to stop such raids. The only effective way to protect captive fish is to string netting above their pools. This is now standard practice in areas where people can afford such measures, but in many parts of the world farmers find nets too expensive to use, especially where ponds are large. As fish farms proliferate in protein–hungry countries, conflicts with Ospreys and other fish–eating birds could well become an issue.

At the other end of the spectrum, I have a neighbor (not a hungry fish farmer) who stocks his backyard ponds each spring for the express purpose of attracting Ospreys. He enjoys watching from the comfort of his breakfast table as neighborhood Ospreys stop by to fish. Few people will go to such lengths to feed Ospreys but, for those who do, the viewing rewards can be great. On a larger scale, stocking a series of ponds with cheap, prolific fish like carp could boost the breeding success of a marginal Osprey population, although it could also damage the ecology of many ponds. A more direct management approach involves feeding the birds at their nests. This has been successful as an experimental technique – Ospreys readily eat fresh fish left at their nests (Poole, 1985) – but it requires too much labor to be useful as a management technique.

10.2 Nest sites

Providing nest sites for Ospreys is much easier than providing them with food, and far more effective. Success is due to two related factors: the scarcity, in many regions, of stable, predator–free natural sites that will accommodate the Osprey's bulky nests, and the intensity of competition for those few good sites that do exist (Chapter 8). Getting around these limitations, the main thrust of Osprey management in recent decades, has meant flooding woodlands to create natural snags for nesting or else building artificial nesting platforms in locations the birds are known to favor.

Beaver (*Castor canadensis*) were probably the first Osprey nest site managers, albeit unwitting ones. The feats of hydrological engineering that these large aquatic rodents can accomplish is truly astonishing. Until fur trappers took their toll, substantial areas of North America were flooded with water held back by Beaver dams, stalwart constructions of sticks and mud. Beaver floodings undoubtedly appealed to Ospreys for the same reasons that artificial impoundments do today: water rising behind a dam kills trees and isolates them from shore, creating safe overwater snags for nesting. In addition, shallow water floodings are often ideal habitat for fishing Ospreys.

In recent decades, hydrologists have taken over where the Beaver left off, and Ospreys have continued to prosper. In western and midwestern North America, for example, the rush to store water for irrigation and waterfowl habitat has been accelerating since the 1920s. Many Ospreys in this region now nest along reservoirs or wildlife floodings, extending the range of the species and building up breeding densities many times higher than along nearby free–flowing rivers (section 3.4).

Although new reservoirs may bristle with natural nest sites, dead trees inevitably decay (Table 8.4). To counter this process and thus to hold local densities of nesting Ospreys, people have built nesting platforms in live trees or on poles in the water (Kahl, 1971; Postupalsky & Stackpole, 1974; Saurola, 1978; Figs. 10.1 and 10.2). These platforms – wooden squares about one meter to a side, which provide a flat stable base for a nest – help to keep Ospreys at established locations and to attract new pairs.

Nest platforms are a management tool with a long history, going back at least to colonial times in New England when farmers placed wagon wheels on poles near barnyards so that Ospreys would nest

Figure 10.1 An Osprey nesting platform built in a topped–off tree, one of several such platforms along the Umpqua River in western Oregon, USA. Ospreys nested and raised young at this platform one year after it was installed. (Photo: Boberg & Witt, US Dept. Interior, Bureau of Land Management.)

and keep the local 'chicken hawks' away (Chapter 11). One of the simplest ways to install a nest platform is to cut the top from a tree (preferably a live one; it will last longer) and then secure a platform to the sawn-off tip (Fig. 10.1). This method eliminates the need for a pole and retains the natural aspects of the surrounding landscape while simulating the flat–topped trees that Ospreys like. The Finns have used tree platforms with particularly good success. Pertii Saurola (1978) and colleagues scaled hundreds of tall trees in Finnish forests, roping up saws and platforms and hammers after them. The idea of tree platforms is also catching on in the western United States and Scotland.

In open fields and marshes, poles or tripods can be used to support Osprey nest platforms. These should be placed well away from surrounding trees and should be protected from predators (Appendix 9). In North America, many people use discarded utility poles

Figure 10.2. In shallow water, tripods provide a stable base for Osprey nesting platforms. This tripod is one of many built to replace decaying nest trees on a reservoir in Michigan (USA). Metal cones around the tripod legs prevent predators from climbing to the nest. See Appendix 9 for a description of how to build these and other artificial nest sites for Ospreys. (Photo: S. Postupalsky.)

as platform supports. Perhaps the solidity and permanence of such structures appeal to the American psyche. New Jersey wildlife personnel carried such nesting poles aloft by helicopter, dropping them like giant darts onto soft marsh peat where they stuck, some at crazy angles. Ospreys used nearly all of them anyway.

In most situations, however, large poles and expensive measures are not really necessary. Shorter lightweight poles, easily installed by two or three people, have proven more than adequate on the open expanses of New England salt marshes (Fig. 10.3). Over swampy ground or shallow water, simple tripods work best to support nesting platforms (Fig. 10.2; Appendix 9). Such overwater sites, isolated and safe from climbing predators, are ideal and the birds compete intensively for them.

Anyone doubting the effectiveness of these artificial nesting platforms should be made to count the Ospreys using them (Table 10.1). Only where platforms are poorly sited, where safer overwater sites (channel markers and duck blinds) are available, or where pairs are already well established at natural sites do Ospreys ignore nesting platforms. In Finland, for example, about 30% of the population nested on tree platforms during the late 1970s, but many platforms went unused because natural (tree) sites are quite stable there (Saurola, 1986). In open locations, by contrast, where natural nest sites are scarcer, I have seen Ospreys circle nesting poles as these were being installed, only to start building nests minutes after the workers departed.

Overall, platforms have accomplished three key management objectives emphasized by Postupalsky (1978):

(1) maintaining pairs in suitable habitat, where natural nest sites might be deteriorating;
(2) maximizing reproductive success and colonization of new habitat, thereby allowing a population to reach the full carrying capacity of its environment; and
(3) facilitating research by making nests easy to reach.

Besides boosting Osprey numbers locally, platforms give people an effective means of luring pairs away from problem nest sites. Ospreys find the high gridwork or double crossarms of utility poles to be irresistible nest sites (Figs. 1.1 and 3.2), but nests there sometimes short out electrical lines. In addition, live electrical lines can be a threat to the Ospreys themselves, especially young fledglings whose clumsy wings are prone to bridge the lethal gap between pole and wire. Judicious use of nesting platforms can help to avoid these problems. In Nova Scotia, when wildlife personnel moved empty Osprey nests from electrical towers to nearby nesting poles, pairs readily shifted over when they returned in spring (Austin-Smith & Rhodenizer, 1983).

Alternatively, wooden utility poles can be modified so that nests do not interfere with power transmission. Bolting a short extension to the top of a pole can support a platform that keeps even a messy

Figure 10.3. The author and friends set up a nesting pole for Ospreys on a New England salt marsh. (Photo: B. Agler.)

Table 10.1. *Regional differences in the structures that US Ospreys chose for nest sites during the late 1970s. Differences among populations reflect differences in availability; the birds favor artificial structures (platforms and pylons) and overwater sites (channel markers and duck blinds) where available.*

	% of population nesting on			
Location	Tree	Channel marker	Duck blind	Nest platform[a]
New Jersey & Delaware	29	1.4	31	38.5
Coastal MD & VA	24	20.5	26	30
Chesapeake	32	28.7	22	18
N Carolina	97	0.8	—	2
N California	92	—	—	8
Oregon	95	0.8	—	4.5
Wisconsin	25	—	—	75
Coastal NY City to Boston	29	1	—	70

[a] Includes nesting platforms, power poles, and towers.
From Henny, Collins, & Diebert (1978a), Eckstein & Vanderschaegen (1988), and Poole & Scheibel (unpublished).

Osprey nest away from the wires below (Fig. 10.4; Olendorff, Miller & Lehman, 1981). Such changes make utility poles good nest sites for Ospreys (a few hundred kilovolts is a very effective deterrent against curious humans and other climbing predators), although birds perching on nearby poles risk electrocution. A second alternative avoids the problem altogether by installing devices that prevent Ospreys from settling on crossarms (VanDaele, VanDaele & Johnson, 1980). All of these modifications require the cooperation of local electric companies, many of which have a long, unselfish history of helping to manage Osprey nest sites.

Ospreys encounter other conflicts trying to adapt to a technological world. In shallow water, channel markers and buoys are favorite nesting sites (Table 10.1; Fig. 10.5). Nests on lighted markers sometimes cause trouble by obscuring lights or covering battery boxes, preventing routine maintenance. Along the east coast of the United States, Coast Guard personnel destroyed Osprey nests on lighted markers, only to reverse this policy during the 1970s thanks to the Osprey's growing popularity and the educational efforts of concerned naturalists. Now the Coast Guard regularly installs alternate lights or battery boxes at nest-laden buoys, an expensive inconvenience. In areas like Chesapeake Bay, where hundreds of nesting Ospreys have settled on navigational markers, it

Figure 10.4. At this electrical pole in southern New England, a utility company built a second set of crossarms above those supporting the wires to keep Osprey nests from shorting out the current. (Photo: A. Poole.)

seems at times as if the Coast Guard had been hired to build fancy nest sites for Ospreys, complete with numbers and blinking lights. How long these good people will continue their liberal policy is anyone's guess. One solution to this problem might be an inexpensive frame nesting platform, such as the one designed by Martin, Mitchell

Figure 10.5. Because Ospreys like to nest over water (doing so provides safety from ground predators), they are strongly attracted to buoys and channel markers. Conflicts arise when these nests obscure numbers and lights or hinder maintenance. (Photo: A. Poole.)

& Hammer (1986), that can be bolted to the side of a buoy away from lights and batteries.

The beauty of Osprey nesting platforms is the flexibility they provide to management efforts. Only a handful of other birds are so easily persuaded to nest in specific locations. Storks nest on chimneys, kestrels and martins use nest boxes – all, like Ospreys, are birds limited in their choice of natural nest sites and quick to lose their fear of man. As a result, nesting poles in the right locations often do attract nesting Ospreys. Not all nest platforms are taken immediately, of course, but in keeping with today's real estate market, waterfront platforms are usually occupied sooner than inland ones. Along parts of the US eastern seaboard, especially Maryland, Florida, and New England, the sight of occupied nesting platforms near docks and houses is not that unusual. Platforms have even become status symbols; party conversations have been known to revolve around the host's Ospreys. Platforms placed near nature centers or other locations with public access contribute to the environmental awareness of a wider community.

Some people object to Osprey nest platforms, seeing them as too manipulative and too ugly. There is no denying the manipulative aspects, but the Ospreys clearly do not mind. The more interesting question is how Osprey platforms affect us – our landscapes and our appreciation of the natural world. Here we enter the realm of aesthetics. Do we really want all Ospreys suburbanized, tamed, and dependent on *Homo sapiens* for nest sites? In some regions we may have no choice. But where environments have escaped relatively unscathed, I believe it is healthy to retain a few local Osprey populations that would never know the difference if plague killed us all tomorrow. Without at least some Ospreys nesting in natural sites, we will never fully understand population regulation in this species, nor appreciate how Ospreys fit into historical landscapes.

By confining Ospreys to artificial sites we are not only taming them to some degree but we are shouldering greater responsibilities than most people realize. Who maintains the platforms? Who decides how large and dense a colony should be – when to stop building platforms in any one area? Who provides effective predator guards for nesting poles? And who keeps the ignorant and the curious from disturbing these sites? These are just a few of the questions that people have ignored in their rush to jump on the nest-platform bandwagon. As more and more Ospreys become platform nesters, these questions will demand some answers.

10.3 Reintroductions

There is a revolution underway among those who manage wild birds of prey. It is a revolution that applies the falconer's time honored technique of 'hacking' to an alarming problem of the twentieth century – the accelerating loss of raptor populations. To hack a hawk is to wean it gradually from human dependence, allowing it to fly free but providing food at a prominent location (the hack site) so the bird's transition to the wild is an easy one. Falconers have been doing this at least since medieval times, usually with the expectation of later recapturing the hacked bird, its hunting skills sharpened by contact with wild prey.

Today, people have modified this technique for conservation purposes. The hawks, gathered as nestlings or born in captivity, are hacked as fledglings (with no expectation of recapture) in areas where populations need restoration. It is assumed that such young will gradually adjust to the wild, migrate successfully, and return to breed in nearby areas, boosting local numbers or recolonizing former range. In several cases, this has actually worked. Managers have achieved notable successes hacking Peregrine Falcons in North America, where a massive reintroduction program is underway involving captive-bred young (Cade, 1982). Populations of White-tailed Sea Eagles in Scotland (Love, 1983) and Bald Eagles in the eastern United States (Nye, 1983) have also benefitted from this technique.

Although Ospreys have been less threatened than these other raptors, hacking projects are catching on quickly with this species. So far, inland regions of the eastern United States have been the focus of all Osprey reintroduction projects (Appendix 10). These are areas where the species was thought to have bred historically, but where it declined or disappeared during the twentieth century. The justification usually given for these efforts, therefore, is that restoring the species to its former range is a worthy goal, especially since Ospreys are slow to disperse and to colonize new territory on their own (Chapter 8).

Providing a nudge in new directions has meant gathering nestlings about six weeks old from viable colonies, loading them into traveling cages on small airplanes, and flying them to hack sites – locations where they can be held, fed, and finally released as fledglings (Fig. 10.6; Schaadt & Rymon, 1983). This is expensive, labor-intensive work. Its success depends on the inherent tendencies of young Ospreys to imprint on the region they see as fledglings (rather than as nestlings), and to return there when old enough to breed.

It is clear now that hacked Ospreys do indeed survive in good numbers and do return to the locales where they fledged. About 20% of the 111 young that Larry Rymon and co–workers hacked near reservoirs in northeastern Pennsylvania had returned to that region by 1986 (Appendix 10). All were identified by colored leg bands. Eleven of these individuals, most of them four-year-olds, paired and bred in 1986, one with an unbanded 'wild' Osprey. These were the first Ospreys known to have nested in this region for decades, and two pairs succeeded in raising young. Even though we await the final verdict on Osprey reintroductions (producing a few young is not the same as sustaining a viable population), the evidence does seem promising. In other regions, Osprey hacking projects have produced less conclusive results, but most are just getting underway (Appendix 10).

There are obvious and important management implications to Rymon's successful hacking project, but the biology revealed here

Figure 10.6 Side view of an Osprey hacking tower in Pennsylvania. Older nestlings, transferred from regions where Ospreys are thriving, are kept in these cages for protection until ready to fly, and are fed and watched through slits in the back wall. Workers swing open the front of the cage when the young are old enough to fly, but continue to provide food at the tower until the young are fully independent.

has been just as interesting in many ways. We see, for example, how critical the post fledging period is for a young Osprey. In these few weeks, a fledgling learns 'home' – where to return after an absence of at least 18 months. We see also that returning young can attract wild birds as mates, even in regions where Ospreys had rarely bred before. Thus reintroduction projects can potentially double their expected returns, although such extra dividends are never assured.

These projects also test the ability of young Ospreys to learn to hunt on their own. Many hacked fledglings caught their own fish only days after first flight (Schaadt & Rymon, 1983). Another finding was that returning males sometimes fed hacked fledglings. It is not clear why these unmated 3–4-year-olds bothered to care for unrelated young, but the behavior is similar to that shown by parent Ospreys that feed fledglings from neighboring nests (Poole, 1982b). As a final note, several of Rymon's hacked young chose to nest in trees when returning as breeders, so hacking at artificial towers does not force later dependence on artificial nest sites.

Are Osprey reintroduction projects really necessary? Ospreys may not colonize new territory quickly, but they have managed to find new breeding sites in the western United States, well outside their former breeding range, as well as in Scotland and southern France where the species had been extirpated for decades (Swenson, 1981; Chapters 3 and 11). Restoration projects, of course, would have introduced Ospreys to these regions much sooner and in much larger numbers. In essence, therefore, hacking lends speed and insurance to the dispersal process.

While it is understandable that people want quick results, some may be jumping on the bandwagon too quickly. A few of the regions chosen for Osprey reintroductions provide marginal nesting and feeding habitat, with only scattered records of the species ever having nested there, probably for good reason. Increasingly, the goal of these projects appears to be introduction, the release of Ospreys into regions where they never bred, rather than reintroduction, the restoration of a damaged population. Without carefully assessing new habitat – sampling fish populations and checking for adequate nest sites – any hacking project is clearly a gamble. Rarely have Osprey restoration projects made such careful assessments.

Perhaps the focus of Osprey restoration projects should be switched to European countries, replacing the current spate of projects in the United States. After all, the US Osprey population is now large, thriving, and well dispersed. Although admirable for their development of methodology, most US hacking projects were set up as much for aesthetic reasons as for the overall health of a regional population.

In Central Europe, by contrast, Ospreys are still struggling, recovering slowly from the effects of persecution (section 3.1). Scandinavia offers a substantial reservoir of nestlings that could serve as hacking stock elsewhere in Europe. The political and social climate in France, the Federal Republic of Germany, Denmark, and perhaps Spain are ripening for such reintroductions, which might generate welcome publicity for Ospreys and duplicate the excitement that greeted their return to Scotland.

While some countries will probably be slow to initiate hacking projects, it is comforting to know that, four or 40 years hence, the potential exists now to redress damage done to Ospreys by earlier generations of men.

11 TWO CASE HISTORIES: SCOTLAND AND NEW ENGLAND

I had often read of the lonely Osprey tenants of one or two silent Scottish lochs, with the watchful eye of a warden constantly upon them, and my surprise can therefore be imagined when I saw my first American Osprey's nest. It was at a popular seaside resort in New Jersey, and perched on a tree overlooking a lake full of row boats and noisy holiday-makers.

C.G. Abbott (1911)

In the spring of 1851, Lewis Dunbar, gentleman naturalist and professional egg collector, carried out one of his many annual raids on Scotland's most famous nineteenth century Osprey eyrie. The eyrie was located atop a ruined castle in picturesque Loch an Eilein, Speyside. As Waterston (1971) retells the story, Dunbar set out that early spring evening and walked about 25 kilometers through a snowstorm, reaching the loch during chilly predawn hours. Undaunted, he waded into the frigid waters, swam to the castle several hundred meters away, scaled the slippery snow-covered ruins, and retrieved two eggs from the nest. On his return trip, he swam on his back with an egg in each hand, but still found enough energy afterward to blow the eggs in a nearby boathouse, washing out the shells with whisky. Even though he was helping to seal the fate of Scotland's dwindling Osprey population, one has to admire Dunbar's toughness and perseverance – to say nothing of his ability to locate whisky in remote areas.

One hundred and ten years after this incident, people were visiting Scotland's most famous twentieth century Osprey eyrie, but for a very different reason. They came in droves to look, to admire a rare and exciting bird, their interest whetted by the 40 years in which nesting Ospreys had been absent from Great Britain. To accommodate these watchers, the Royal Society for the Protection of Birds (RSPB) built an observation blind (hide) in 1959 at Loch Garten, where the returning Ospreys had first settled. Thousands of visitors arrived in that first year. Twenty-two years later, the millionth visitor walked up the pine fringed path to the Loch Garten blind and was presented, much to her surprise, with a specially engraved glass bowl donated by a local business, Osprey Electronics. Ospreys, in short, had become a major tourist attraction in Scotland by the 1980s.

At the same time that Ospreys were recolonizing Scotland, they were losing their grip in New England (USA). Contamination, rather than persecution, was the problem here. Studies of New England Ospreys during the 1960s showed high levels of DDT and other contaminants in eggs, pitifully low breeding rates, and plummeting numbers of breeders. Roger Tory Peterson (1969), the celebrated naturalist, author, and artist, was among the first to sound the alarm. Osprey numbers near his Connecticut home dropped from 150 nesting pairs in 1954 to 10 pairs in 1968, and he predicted these birds would soon disappear from the northeastern United States. His prediction might well have come true, but unforeseen factors – insect resistance, public outcry about toxins in food and the environment, and responsible legal and government action – curbed the flood of pesticides in this region. As the use and impacts of toxins eased, Osprey eggs started hatching again. This southern New England population now gives hints of reaching its former abundance.

It is the Osprey's history in Scotland and New England, the shifts in numbers and the human attitudes precipitating those shifts, that concerns us in this chapter. The focus is on these two histories because they have been so well documented. No other Osprey population has coexisted with people who were as interested in its fate and as eager to record it. In tracing these histories, we glimpse many of the themes that have emerged in previous chapters, especially those chapters dealing with threats, management, and population dynamics. But here the themes are elaborated with the human element given particular emphasis, for it is impossible to record the history of Ospreys in Scotland and New England while ignoring their human neighbors. The fate of these birds has been linked increasingly with the interest and goodwill of humans. As our planet's human population doubles in the next 20 to 30 years, the same will undoubtedly prove true for Ospreys around the globe.

11.1 The British Isles

Decline and extinction

Scattered anecdotal records of Ospreys nesting in Great Britain date back at least to 1800, perhaps earlier if Gaelic place names are any indication. 'Isgair' or 'Iasgair' (the fisherman) are names used even to this day for several lochs in the Scottish Highlands, a reference to the Ospreys that must once have bred in those locales. But accurate records of Osprey numbers and distribution in the British Isles emerged only during the mid-1800s. By then, British birds of prey had been heavily persecuted, so reconstructing details of earlier populations was difficult. At the most, perhaps 40–50 active Osprey nests remained in 1850, all confined to central and western portions of the Scottish Highlands (Waterston, 1971; Bijleveld, 1974). A few scattered pairs had also nested in England (Somerset) as late as 1842 and in Ireland (County Leitrim and near Lake Killarney) up to about 1800, so Ospreys once had a wide, if patchy, distribution in Great Britain (Blathwayt, 1921; Sandeman, 1957).

Persecution during the seventeenth and early eighteenth centuries undoubtedly contributed to Osprey nest scarcity in Scotland, but there are few records to prove this. By the 1840s, however, it was clear that persecution was taking a toll and that the Osprey's rarity was accelerating its demise. As nest numbers dwindled, eggs and skins became increasingly sought after – eggs to decorate glass cases and felt-lined drawers of collectors, stuffed and mounted skins to preside over the draughty halls of country manor houses. Lewis Dunbar carried out five successive raids on the Loch an Eilein nest between 1848 and 1853, robbing the eggs each time. From 1846 to 1900, the fate of 24 breeding attempts were recorded at this nest: 13 robberies, one shooting, and two desertions apparently caused by disturbance (Cash, 1914; Waterston, 1971). If this was typical of the level of persecution at other Scottish sites, then late nineteenth century pairs succeeded in raising young only about once in every three times they bred. Assuming an average brood of two young, this meant that only about 0.66 young fledged per nest per year, clearly too few to balance adult mortality even under the best of circumstances (Table 8.2). Little wonder that so many of Scotland's nineteenth century Ospreys nested on islands in lochs, probably gaining some protection by the surrounding water from casual persecution (Fig. 11.1).

Yet persecution cannot have been uniformly intense because Scottish Ospreys faded slowly. At least a few breeding pairs lingered on into the early twentieth century (Sandeman, 1957). Assuming (generously) that Scotland supported 50 breeding pairs in 1850, this population lost less than one pair per year, on average, between 1850 and 1916. While some nests were being raided relentlessly, therefore, others probably managed to fledge young regularly. This success is all the more remarkable considering that people regularly shot these birds. Because Ospreys readily defend their nests, often circling overhead when disturbed, nineteenth century sportsmen like Charles St John and William Dunbar (Lewis's brother) found them easy targets. Here is how St John (1849) describes one of his many encounters with the species:

> On coming nearer, we could distinguish the female Osprey on the nest. It was determined that I should remain concealed near the loch while my two companions went for the boat . . . that I might have the chance of shooting the old Osprey herself in case she came within shot. I must say that I would rather she had escaped this fate; but as her skin was wanted, I agreed to try to kill her . . . She passed two or three times, not very far from me, before . . . at last I fired and the poor bird, after wheeling around for a few moments, fell far to leeward of me, and down amongst the most precipitous and rocky part of the mountain, quite dead.

Even without the hyperbole, such killing is easy to condemn from today's perspective. But these men were typical naturalists of their day; nineteenth century field ornithology was commonly practiced with a gun. Modern binoculars and conservation ethics have changed

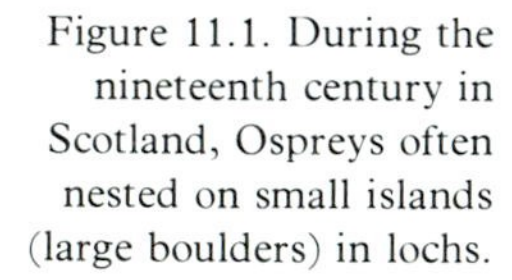

Figure 11.1. During the nineteenth century in Scotland, Ospreys often nested on small islands (large boulders) in lochs.

our outlook, but I suspect that many of today's naturalists, had they been born 150 years ago, would have eagerly joined St John's adventures. A few Scottish landowners, to be sure, did protect Ospreys in their districts, but these were farsighted exceptions (Cash, 1914).

It is more difficult to forgive the hunters their hypocrisy, however. At the same time that he was blasting Ospreys from the skies, St John (1849) wrote:

> Why the poor Osprey should be persecuted I know not, as it is quite harmless, living wholly on fish, of which everyone knows there is too great an abundance in this country for even the most rigid preserver to grudge this picturesque bird its share. The fact probably is that his skin is worth something to keepers and others, as they can always get a few shillings for it, and therefore the bird is doomed to destruction.

It is not clear whether sportsmen like St John or the local Highland gamekeepers that he blames did the most harm to Scottish Ospreys. Undoubtedly both were to blame. The last pair known to nest in Great Britain (until recently) was found on an island in Loch Loyne in 1916. As William Dunbar wrote to his fellow egg raider John Wolley: 'I am afraid that Mr St John, yourself, and your humble servant, have finally done for the Ospreys.'

Recolonization and recovery

Details of this recovery period are provided by Waterston (1971). Single Ospreys were seen occasionally in Scotland from 1920 to 1950, mostly during spring and fall migration, but none was known to nest anywhere in Great Britain during those years. Presumably most of these birds were migrants from nearby Scandinavia. Two Ospreys recovered in the Scottish Highlands in 1949 and 1955 had been banded as nestlings in Swedish eyries. During the early 1950s, a few such migrants began lingering on through the summer. By the mid 1950s there were scattered reports, none substantiated, of Ospreys breeding in Scotland: a nest built at Loch Garten in 1955 and one in the nearby Rothiemurchus Forest, Speyside, in 1956. (The 1950s were a decade of rapid growth in the Scandinavian Osprey population, due to increased protection (section 3.1); this probably explains why Scotland was recolonized during those years.) Even though these pioneers generated great excitement among conservationists, persecution continued, albeit sporadically. One Osprey was shot at Loch Garten in 1956 and the following year a thief eluded guards one dark, drizzly night and made off with a clutch of eggs.

That robbery galvanized efforts to protect nesting Ospreys in Scotland. Hundreds of acres surrounding the nesting tree at Loch Garten were set aside as a sanctuary and the RSPB stepped up nest surveillance. When an appeal for volunteer wardens went out, the response was overwhelming. To accommodate these new guards and their support staff, an Osprey camp was established at Loch Garten in 1958, complete with tents, cooking facilities, a small guard hut overlooking the nest site, tangles of barbed wire surrounding the base of the nest tree, and a sensitive nighttime microphone to monitor movement near the nest. The camp looked for all the world like a commando training post.

But these efforts eventually paid off. After another robbery in 1958, three chicks hatched in the Loch Garten nest the following year, the first Ospreys known to have hatched in Scotland in over 40 years. The news traveled fast. Newspaper reports of the 1958 robbery had already made the Loch Garten pair famous, and with the added lure of nestlings, people poured in for a look. Nearly 14 000 visitors arrived in that first year of hatching, a tribute to the broad network of naturalists in Britain, and an observation post was built to accommodate them (Fig. 11.2). By channeling access to a single nest

Figure 11.2. The RSPB observation post at Loch Garten, Scotland. This blind can accommodate at least 20 viewers and is equipped with telescopes and binoculars which allow visitors to watch the nearby pair of nesting Ospreys without disturbing the birds. (Photo: A. Poole.)

site, and later to a second one at the Scottish Wildlife Trust near Perth, the RSPB had solved a major problem – disturbance of nesting Ospreys by egg robbers and by over-eager birders and photographers. Guaranteeing public access to two nests meant that the locations of others could be kept hidden. Even today, nests are robbed. Five of 29 Osprey nests active in Scotland during 1984 lost eggs to vandals, and rarely a year goes by without a robbery at one or two sites – despite a potential fine of £2000 (R. Dennis, unpublished).

It seems paradoxical that citizens of the United Kingdom, one of the best educated, best fed, and most conservation–minded nations on earth, should continue to steal eggs from rare birds. Most of today's nest robbers, however, seem to have different motives than their predecessors did. While a few Osprey eggs are still being robbed for collections, most are taken as a challenge, perhaps just for the publicity of such a prank. Reports from northern England, for example, tell of vandals displaying their recently stolen Osprey eggs in pubs, while friends applaud. Since many of Scotland's Ospreys nest on large private estates, class resentment may partly motivate the robberies.

Annoying as this recent nest robbing has been, its impact at the population level has not been significant. Sporadic nest robbing functions much like bad weather, temporarily reducing reproductive output locally, but rarely affecting long term population trends. Since 1970, for example, Scottish Ospreys have produced, on average, about 1.6 young per active nest, high breeding rates by any region's standards (Dennis, 1983; Appendix 4). Growth of Scotland's Osprey population has reflected this output, with the number of breeding pairs doubling every five to seven years (Fig. 11.3, bottom). Such rapid growth rates, as we have seen, are typical of areas where populations are small in relation to available resources, especially food and nest sites.

So far, Osprey recolonization of Scotland has been restricted to the northeastern Highlands. A favored area has been the districts of Badenoch and Strathspey, a rich mosaic of birch and pine woods, croftlands, lochs, and marshes, all drained by the meandering river Spey – famous for salmon and whisky. The lowland lochs and forests here are reminiscent of Osprey habitat in southern Scandinavia, perhaps one reason that Scandinavian migrants first recolonized this area. Productive lochs, many stocked with fish, have been key attractions for the growing population. In addition, numerous trout ponds and hatcheries dot the region, a predictable supply of prey. Even though protective netting has now excluded Ospreys from most Highland hatcheries, trout from ponds and lochs remain a key part of the Scottish Osprey's diet.

When a BBC crew wished to film the Loch Garten Ospreys fishing, they set up a blind alongside a local trout pond, where Ospreys were regular visitors. When an Osprey appeared, one of the crew would toss a few pellets of trout food on the water, trout would inevitably rise to the bait, and the Osprey would just as inevitably spot the fish and plunge. By tossing the bait to different parts of the pond, the crew could position the Osprey's dive exactly where the cameramen wanted it.

There are many regions of Great Britain that Ospreys have not yet recolonized, even though habitat appears suitable. The western Highlands, for example, once a stronghold for the species, remain

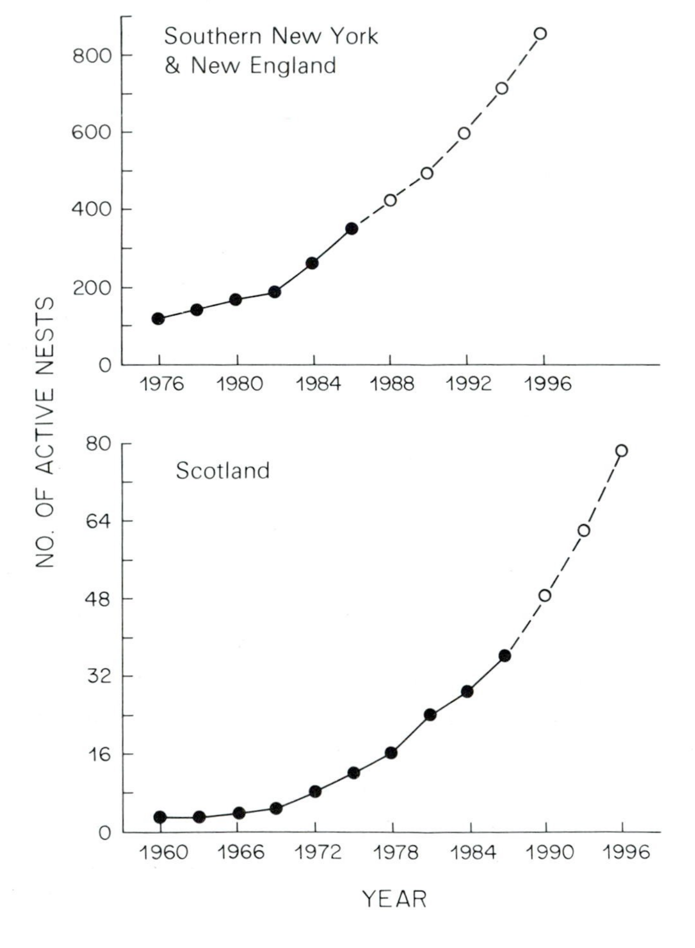

Figure 11.3. Growth of Osprey populations in southern New York (Long Island) and New England and in Scotland during recent decades. Dashed lines and open circles show projected growth for these populations, assuming current trends continue. Scotland from Dennis (1983 & unpublished); New York–New England from Spitzer (1980), A. Poole (unpublished), and various state wildlife agencies in the region.

uncolonized. Likewise uncolonized is Britain's coast, even though Ospreys reach their highest breeding densities along coastal estuaries in other parts of the world. A few pairs, to be sure, do nest near the Firth of Forth, feeding in local estuaries, but most of Scotland's Ospreys are inland breeders dependent on lochs, rivers and ponds.

The future

Current trends suggest that at least 100 pairs of Ospreys will nest in Great Britain by the year 2000 (Fig. 11.3). Food and nest sites appear adequate, and serious threats are lacking, so far. Contamination of Scottish Ospreys by pesticides, for example, has always been low (R. Dennis, unpublished). It is nest site availability that will probably determine when Osprey numbers start leveling off here. About 30% of Scotland's Ospreys now nest on artificial platforms, most of them built in treetops by RSPB personnel – an effort to retain natural looking nest sites. (Modifying electrical pylons to support Osprey nests might also attract new breeders and protect nests. No sane egg robber would risk climbing near high voltage lines.) If such limited nest site management continues, and it undoubtedly will, then the British Isles might eventually support several hundred breeding pairs, perhaps even a few small colonies such as those in East Germany (Fig. 3.2).

Dispersal remains the biggest unknown in the future of Britain's Ospreys, although dispersal away from current concentrations in Scotland is likely to be slow. Empty nest platforms near centers of Osprey concentration could actually hold potential dispersers, preventing the colonization of new and distant areas (Chapter 8). Hacking (section 10.3) is one way to speed this dispersal. Nestlings could be taken from thriving populations in Scandinavia and released at hack sites in appropriate areas of Great Britain. Because hacking has proven so very successful in re–establishing populations of breeding Ospreys in the United States, this is clearly the way to reintroduce the species to its former haunts in England and Ireland.

The spread of Ospreys through the British Isles would be a healthy change. Not only would it broaden this population's base, reducing vulnerability to local shifts in weather, contamination, and food supply, but it might also help to reduce the threat of nest robbing. The eggs of common species are seldom sought avidly. Even if sporadic robbing continues, a larger, more dispersed Osprey population would better absorb the loss. Perhaps, too, new studies will be encouraged as Britain's Ospreys became more numerous and better distributed. So far, this small population has been carefully guarded, discouraging research.

The symbiosis that has sprung up in Britain between Ospreys and people is truly remarkable. People provide Ospreys with protection and nest sites; Ospreys provide people with the excitement of a wild, rare predator that can be viewed at close range. The Scottish countryside has been given a spark of new life. British newspapers and television carry reports each year of the Osprey's arrival and breeding success: this bird is now headline material. Such attention is all the more remarkable considering how eager an earlier generation was to persecute it.

Other birds, of course, have also precipitated such quick shifts in attitude, but none has done this so emphatically, no other has generated as many guardians. Except for East Africa's wildlife, North America's whales, and perhaps China's Panda, no other wild animal has attracted such devoted attention. As they bask in their glory, British Ospreys of the future will be more numerous, more visible, more widespread, and (perhaps) less preciously guarded than they are today.

11.2 Southern New England and Long Island, NY *1840–1940*

Nineteenth and early twentieth century Ospreys were probably just as heavily persecuted in the northeastern United States as in Scotland, but there were many more US Ospreys to absorb the losses. Exactly how many more remains open to question, however. In southeastern New York State (Long Island) and southern New England, the well–studied region that concerns us here, no accurate surveys of Ospreys were made before the 1930s. At that time, people estimated about 800 to 1000 breeding pairs, nearly all of them nesting along the coastline between New York City and Boston (Spitzer, 1980; Poole & Spitzer, 1983). Perhaps 1500–2000 pairs had nested in this region a century earlier.

Shooting, habitat destruction, and to a lesser extent nest robbing all took a toll on these birds. In 1882, J.N. Clark (unpublished), a naturalist and collector for the US National Museum, wrote in a letter as follows about the Ospreys near his seaside home in Saybrook, Connecticut:

> Forty years ago there were a number of fish hawks nesting within our limits and, although all those of easy access were annually

> robbed, there were a few trees inaccessible to the small boy where the young were reared in peace. But with the advent of summer boarders and seaside excursions, sports with the best modern shooting appliances found it no difficult matter to shoot the parents . . . and I do not think there is now a nest within our limits.

Although other Ospreys in the New York–New England region were also persecuted during those years, many were carefully protected. On Plum Island, NY, a colony of several hundred nesting pairs had sprung up during the 1800s, thanks in part to protection by a local lighthouse keeper. Most of these birds, however, deserted after hotels were built and summer tourists arrived with guns (Allen, 1892). In its heyday, the Plum Island colony was probably the largest Osprey nesting colony in the world, a good example of how protection, abundant food and nest sites, and the absence of predatory mammals encourage high breeding density in this species. C.S. Allen (1892), an American ornithologist and egg collector, visited Plum Island in 1879 and 1885, before development. With 'good cigars and a thorough appreciation of his pets (the Ospreys),' Allen persuaded the lighthouse keeper to provide tours of the colony. Here is some of what he saw:

> On nearing the island . . . one was struck with the great number of fish hawks to be seen on all sides . . . sailing through the air or perched on the stakes of fish ponds . . . The first nest shown me was in a dooryard only about 50 yards from a house and only seven or eight feet from the ground . . . In the wooded part of the island the nests were very numerous, the larger trees in the interior all being occupied, while near the edge of the wood nearly every tree had a nest and some of them two or three each . . . On the north shore, where the beach is strewn with large boulders, nearly every rock – even some that were far out in the water – was occupied with a small nest.

At the same time that development was forcing Ospreys off Plum Island, nearby Gardiner's Island – ecologically similar but larger, more diverse, and totally wild – supported another huge colony. Plum's Ospreys were thought to have deserted to Gardiner's, but there was never any proof of such dispersal. Ospreys on Gardiner's had enjoyed decades of protection, so the island probably had little room for new pairs. In any case, ornithologists that visited Gardiner's in the early 1900s found a remarkable concentration of Ospreys there, estimated at 200–300 pairs (Chapman, 1908; Abbott, 1911; Knight, 1932). Nests were everywhere: on meadow stumps, in forest trees, on boulders in fields and along the shore, on fences (Fig. 11.4) and stone walls, even on fish boxes washed up along the beach (Abbott, 1911). Along one short stretch of beach, there was a succession of no less than 22 nests at intervals of from ten to 200 meters. The island's wealthy owners protected these birds, as did the local fishermen who camped there, even though Ospreys occasionally snatched fish from their traps.

There were other protected concentrations of nesting Ospreys in the New York–New England region during the late nineteenth and early twentieth centuries. One of the largest had formed near the mouth of the Connecticut River where almost 200 active nests were clustered on salt marsh islands and along the edges of surrounding forests. Farther east, along the coast of Rhode Island and southeastern Massachusetts, smaller, more scattered breeding colonies had sprung up. This was one of the first areas where Ospreys nested on artificial platforms. E.H. Forbush (1927) tells why farmers were especially keen to have nesting Ospreys as neighbors:

> Here and there someone has erected a tall pole in the dooryard with a cartwheel fixed horizontally across its top. This makes a convenient and safe location of which the Osprey is not slow to

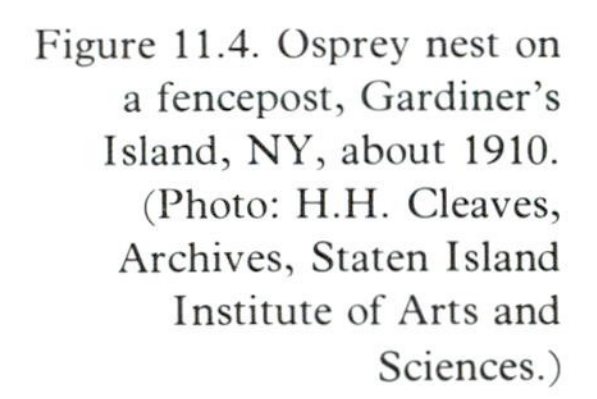

Figure 11.4. Osprey nest on a fencepost, Gardiner's Island, NY, about 1910. (Photo: H.H. Cleaves, Archives, Staten Island Institute of Arts and Sciences.)

> take advantage . . . It seems that while these birds are incubating and rearing their young, they will not allow other birds in the vicinity of their nests, and, as the young chickens are allowed to run at large at that season, the Ospreys protect the chickens from the forays of other hawks.

These cartwheels, of course, were the forerunners of today's nesting platforms. Apparently Ospreys were just as anxious to find safe, solid nesting sites in the nineteenth century as they are today. That sometimes meant nesting atop buildings (Fig. 11.5).

Decline and recovery

People noted declines in Osprey success and breeding numbers in a variety of coastal regions between New York City and Boston during the 1950s, but the cause of this failure was first identified in Connecticut. There, at the mouth of the Connecticut River, Yale graduate student Peter Ames began surveys in 1957 and found Osprey numbers declining at the alarming rate of about 30% annually (Ames & Mersereau, 1964). Thinking that predation might be the problem (many pairs nested low to the ground on marshy islands), Ames joined forces with Roger and Barbara Peterson and built numerous predator–proof nesting platforms for the birds. Pairs readily accepted the platforms but still reproduced poorly, at only about 15% normal levels for the region. These findings prompted the studies discussed in section 9.6: Ames's (1966) egg and fish analyses and Spitzer's (1978) transfer of eggs to Connecticut from less polluted areas. Together these studies left little doubt why local Ospreys were doing so poorly: DDT and other organochlorines, ingested in food, were contaminating eggs and preventing them from hatching.

Studies in nearby regions showed that this problem was not restricted to Connecticut (Wiemeyer *et al.*, 1975; Spitzer *et al.*, 1977; Puleston, 1977). Overall, 90% of the Ospreys nesting between New York City and Boston disappeared between 1950 and 1975 (Table 11.1). Although this was a time of great pessimism – 'progress' had, as usual, brought unforeseen ecological complications – this was also a time when efforts to protect the environment gained momentum. Alarmed that Ospreys and other local wildlife were accumulating DDT, a group of lawyers brought suit against the Long Island Mosquito Control Commission to block continued use of DDT on eastern Long Island. Encouraged by quick success in that suit, and seeing the need for continued public action, these lawyers formed the

Figure 11.5. Osprey nest on a schoolhouse belfry, Cape May County, NJ (USA), about 1910. (Photo: H.H. Cleaves, Archives, Staten Island Institute of Arts and Sciences.)

Table 11.1. *Decline and growth in the number of Ospreys nesting between New York City and Boston, 1940–1986.*

Area	No. of active nests in 1940	1970	1986
Gardiner's Island, NY	300	38	48
'North Fork,' Long Island	79	10	30
'South Fork,' Long Island	68	10	23
Shelter Island, NY	41	16	21
Connecticut River estuary and environs	200 +	8	28
Rhode Island	120	8	25

From Spitzer (1980) and M. Scheibel, L. Suprock & J. Victoria (unpublished).

Environmental Defense Fund (EDF), a public interest law group specializing in environmental concerns.

Seeking a nationwide ban on DDT, EDF and other citizens' groups helped to establish precedents for what we now call environmental law. One key precedent gave legal standing to citizens' groups. Before this, only people with an economic interest were allowed to bring suit in such cases. But the public, it was argued, ate the same fish that Ospreys did and thus accumulated the same toxins, many of them suspected carcinogens. This critical link between people and wildlife, their dependence on a common environment, was finally gaining broad public attention.

Another response to declining Osprey numbers on Long Island and in southern New England came not from lawyers but from naturalists. The Osprey's readiness to accept nesting platforms encouraged state agencies and local naturalists to erect dozens of new structures, far more than the struggling population could use right away. But as DDT residues faded from the environment and Osprey young began fledging again, new breeders flocked to these vacant platforms. Six out of every seven new pairs recruited between 1975 and 1986 have nested on platforms or other artificial sites, and areas with the most platforms have attracted the most new breeders (Table 8.5).

By the 1980s, Osprey platforms were making newspaper headlines. Hundreds of people wanted them. On the island of Martha's Vineyard, Massachusetts, a well–known resort for the rich and the famous, residents pay $350 to have a nesting pole installed on their property. A local naturalist specializes in the task. So many people have asked for poles that there is now a long waiting list. Jackie Onassis has one. So does singer Livingston Taylor. His brother, James, does not have his yet, but he is on the waiting list. All of this activity has accommodated about 45 pairs currently nesting on the island. Other coastal communities between New York City and Boston have experienced a similar Osprey craze, something that coastal residents of Maine, Florida, and Maryland have lived with for a decade or two (Chapter 10).

With troublesome contaminants fading from their environment, and safe, new nesting sites proliferating, the Osprey's recovery in the New York–New England region has been nothing short of remarkable. Since the mid-1970s, the low point for this population, breeding numbers have grown about 10% each year through 1984, and about 20% annually between 1984 and 1986, thus doubling in size every six to seven years (Figs. 8.4 and 11.3). Scotland's Osprey population, which also had access to plentiful food and nest sites, grew at about the same rate during those years.

Nearly all new pairs in southern New England and eastern Long Island have nested along the coast. Plentiful food and nest sites seem to be the attraction. No Osprey in this region is known to feed exclusively in freshwater lakes or ponds. Nesting distribution is roughly similar to that seen historically, except for the current absence of breeders along Narragansett Bay, once a stronghold for the species. In addition, platforms have lured a few pairs to localities where Ospreys never bred historically, most notably windswept Nantucket Island, 30 kilometers south of the New England coast and about 20 kilometers south of Martha's Vineyard.

The future

Along a three-kilometer stretch of the Massachusetts coastline (the Westport River estuaries), I visited over 50 active Osprey nests in a single day last week, twice as many as I censused there five years ago and almost three times the number seen there during the nineteenth century. Safe nesting platforms have been the key to this colony's explosive growth. A decade from now, if current growth rates continue, this colony will hold at least 150 breeding pairs, and the larger region of southern New York and New England will hold at least 800 pairs (Fig. 11.3). How long can such extraordinary growth rates be sustained?

If the current craze for Osprey nesting platforms continues, and it probably will, plenty of safe, vacant nest sites will be available for pairs in the years ahead. Food, not nest sites, will become the limiting factor, although no one knows when that food limit will be reached. New England Ospreys depend on fish that are resilient, abundant, and migrate seasonally to coastal waters from offshore. Unless human fisheries or pollution take a toll on this resource, Ospreys themselves should have a limited impact. A few Ospreys fishing each local bay or estuary for half the year will rarely render migratory fish so scarce that nestling Ospreys starve. But a few dozen Ospreys nesting near each local bay and estuary will add up to a significant population. About 1000–2000 pairs of Ospreys nested between New York City and Boston during the nineteenth century. I think we can expect at least that many in the decades ahead, probably more.

Although Ospreys in the New York–New England region may soon be as abundant as they were historically, they will no longer be the rural and semi–wilderness population that nested there before the 1960s. The region's coastline is fast becoming suburbia; development pressures are intense. But Ospreys are adjusting, as we have seen. They proved to be excellent indicators of environmental contamination but poor indicators of changes that come with development: land clearing, disturbance, and nutrient overloading of coastal waters.

What we are gaining as this region's Ospreys become tamed and suburbanized is the chance to live side by side with a magnificent bird of prey. Clammers, fishermen, vacationers, boaters, pierhead loafers, shoreline homeowners – all are becoming acquainted (some reacquainted) with Ospreys. Some of these people will undoubtedly become knowledgeable about the species. By forging such links, however tenuous, people expand their horizons and gain new awareness of the world beyond their own backyard. Ospreys plunging for fish or disappearing each fall for distant wintering grounds are a strong reminder that no wild animal can be studied in isolation. South American land use policies, Caribbean hurricanes, local fisheries regulations and controls on pollution – all will affect the number of Ospreys that return each spring to New England.

What we are losing as these Ospreys become suburbanized is the chance to see them in their natural setting, the setting in which they evolved. Only a few islands along the coast of Long Island and southern New England will provide wilderness nesting preserves in the twenty-first century. Two or three decades from now, with luck, people may visit these preserves and glimpse again the huge, dense nesting colonies that naturalist C.S. Allen (1982) wrote about during the nineteenth century, colonies in which forest trees and open beaches were strewn with thriving Osprey nests. That will be a legacy worth handing on to the next generation.

In the quote that opens this chapter, a British naturalist who visited the United States at the turn of the century was surprised to find an American Osprey nest 'perched on a tree overlooking a lake full of row boats and noisy holiday-makers.' He had known only the 'lonely Osprey tenants of . . . silent Scottish lochs.' British naturalists who visit the northeast coast of the United States today will find that same contrast. Scotland's Ospreys remain largely a wilderness or rural population, while most New York–New England Ospreys nest within sight of a house, a busy harbor, or a highway. It is this adaptability that defines the species. Trying to grasp the Osprey's essential nature is like confronting the Greek god Proteus, who could change his shape at will. Ospreys are guardians not just of wilderness lakes but also of commercial harbors, desert islands, salt marshes, mangrove estuaries, rainforest rivers, and a host of other ecosystems. All of these varied habitats give life to Ospreys and Ospreys, in turn, give life to them.

APPENDICES

APPENDIX 1.

Estimate of the world population of breeding Ospreys in 1985, subdivided by region.

Regions differ greatly in the degree to which they have been surveyed in recent years. Estimates are presented here in descending order of accuracy. Estimates without question marks are based on surveys in most (> 80%) of a population's known breeding range. Estimates with one question mark are extrapolations from surveys in small portions of a population's range; similar densities were assumed in unsurveyed parts of these ranges. Numbers with two question marks are tentative guesses based on subjective descriptions of Osprey abundance. Sources (references) without dates have not published their data.

Region	No. of breeding pairs	Source
Britain	45	R. Dennis
USA (contiguous states)	7500–8000	Henny, 1983
Mexico (Baja and Gulf of California)	800	Henny & Anderson, 1979
Mediterranean, Spain & Portugal	45–55	Witt *et al.*, 1983; Berthon & Berthon, 1984; L. Gonzalez
Central Europe:		
Poland	20–30	W. Krol
German Democratic Republic (E Germany)	110	Meyburg & Meyburg, 1987
Sinai	45	Y. Leshem
Scandinavia:		
Sweden	1800–2000	Österlöf, 1973;
Finland	900–1000	Saurola, 1986
Norway	150–200	A. Haga
Atlantic Is.:		
Canaries	10–15	L. Gonzalez

Appendix 1 (*cont.*)

Region	No. of breeding pairs	Source
Cape Verdes	50?	de Naurois, 1969
Canada & Alaska	10 000–12 000?	Gerrard *et al.*, 1976; Wetmore & Gillespie, 1976; Prevost *et al.*, 1979; Bider & Bird, 1983; Stocek & Pearce, 1983; Hughes, 1986.
Caribbean (Cuba, Bahamas, & Belize)	100–150?	Sprunt, 1977; Wiley, 1984; Wotzkow, 1985.
S Red Sea and Arabian Peninsula	40??	Clapham, 1964; Österlöf, 1965; Gallagher & Woodcock, 1980.
Australia	200–400??	Blakers *et al.* 1984
Japan	20–50??	Anon. 1974, S. Hanawa, M. Brazil.
USSR	2000–6000??	Galushin, 1977; Galushin & Flint, 1983; Lobov, 1985.
TOTAL	23 835–30 990	

APPENDIX 2.
Temperature data, during the coldest month of the year, for selected regions that support migratory or nonmigratory Osprey populations.

Frost-free period is the longest continuous period of the year without freezing temperatures. Data, from Landsberg (1970, 1971, 1974), are based on 10–30 year norms.

	Mean no. days with temperature below freezing	Frost-free Period (days)
Regions with resident Osprey populations:		
Australia	1–3	300–365
Western Mediterranean	2–4	300–350
USA (Florida)	1–4	280–340
Regions with migratory Osprey populations:		
USA:		
(Georgia)	14	270
(N Carolina)	15	240
(Maryland)	25	210
(Massachusetts)	25	180
Europe:		
(Scotland)	23	245
(Sweden)	28	200
(Germany)	24	210

APPENDIX 3.

Daily energy budgets calculated for an 'average' male Osprey in two different situations: when feeding a family on breeding grounds and when alone on wintering grounds.

For the breeder, data were gathered at three nests in Massachusetts, each watched for 30 hours; each male fed himself plus his mate and three young about 20–30 days old (Poole, 1984). Wintering individuals were observed along the coast of Senegal, West Africa (Prevost, 1982). Energy expenditures were first calculated as multiples of the Osprey's resting metabolic rate (SMR) and then converted to kilocalories via formulas appropriate to a bird of that weight and species (see Prevost, 1982; Wasser, 1986).

	Breeding	Wintering
Energy intake per day:		
number of fish	6–8	1–3
daily catch (grams)	1250	300–350
male's share (grams)	400	300–350
male's share (kilocalories)	360	200–250
Energy expended per day (kcals):		
hunting & flying	220	32
resting & sleeping	117	132
Total	337	164
Hunting efficiency (kcals per min hunting)	5.5	6.1
Minutes to meet daily requirements	195	30

APPENDIX 4.

Regional differences in Osprey clutch size and reproductive success.

Eggs and young 'lost' include both those that disappeared and those found dead. *N* is the number of nests sampled.

Region (Years)	Clutch size $\bar{x}$ (mode)	% Eggs lost	No. young hatched/nest	% Young lost	No. young fledged/nest	% Nests failed	Study (*N*)
Massachusetts, USA (1979–83) C[a]M[b]	3.3 (3)	29%	2.35	19%	1.92	11%	Poole, 1984 (94)
New York, USA (1978–79) CM	3.2 (3)	68%	1.03	20%	0.82	53%	Poole, 1982a (110)
Florida, USA (1978–79) CR[c]	2.7 (3)	64%	0.97	46%	0.52	54%	Poole, 1982a (48)
Scotland (1954–81) F[d]M	2.8 (3)	40%	1.68	7%	1.56	25%	Dennis, 1983 (191)
Baja, California (1977–78) CR	2.8 (3)	46%	1.50	36%	0.97	27%	Judge, 1983 (67)
Chesapeake Bay, USA (1972–74) CM	2.9 (3)	48%	1.52	15%	1.29	31%	Reese, 1977 (327)
Sweden (1971–73) FM	2.8? (3?)	—	—	18%	1.48	26%	Odsjö & Sondell, 1976 (96)
Corsica (1977–79) CR	2.7 (3)	21%	2.14	12%	1.88	8%	Bouvet & Thibault, 1980 (25)
Michigan, USA (1982–85) FM	2.98 (3)	32%	2.02	17%	1.68[e]	42%	Postupalsky, 1985 (51)

[a] coastal population; [b] migratory population; [c] resident population; [d] fresh-water population; [e] based on occupied nests.

APPENDIX 5.

Success of Osprey clutches containing three as against four eggs, over a five-year period.

Data from south coastal New England, USA (Poole, 1984 & unpublished). Eggs and chicks 'lost' were those that died or disappeared.

	3 eggs	4 eggs
No. of clutches	110	53
Full clutch hatched	41%	36%
Full clutch fledged	37%	10%
Zero young fledged	16%	11%
No. eggs lost/clutch (%)	1.0 (34%)	1.24 (31%)
No. chicks lost/clutch (%)	0.44 (14%)	0.76 (19%)
No. young hatched/clutch	2.0	2.76
No. young fledged/clutch	1.56	2.0

APPENDIX 6.

Loss of eggs and chicks from Osprey nests, in relation to laying dates.

Data (from Poole, 1984) were gathered in Westport, Massachusetts during 1980–1983. *N* is the number of nests sampled.

	Laying date (No. days after 1st egg in colony)			
	1–8	9–16	17–24	25+
Eggs				
No. dead/nest	0.68	0.84	1.27	1.85
% dead	20%	26%	40%	59%
% nest with all eggs dead or lost	0%	3%	29%	36%
N	34	32	17	14
Chicks				
No. dead/nest	0.26	0.41	0.31	0.75
% failing	10%	17%	16%	58%
% nests with all chicks lost	0%	0%	7%	12%
N	34	31	13	8

APPENDIX 7.

Levels of DDT (ppm, wet weight) in Osprey eggs from different regions of the northern hemisphere; levels shown include DDT plus derivatives (DDD, DDE) in addled eggs, unless otherwise noted.

Accompanying data on shell thinning (% decline from pre-DDT norms), reproductive success (young/nest with eggs), and population status (+ = stable or increasing; – = declining) show levels of DDT that have affected Osprey status and breeding in different regions. ND = no data.

Region (years)	DDT mean (range)	Shell thickness	Young/ nest	Status	Study
Sweden (1971–73)	4.3 (0.2–32)	–11%	1.5	+	Odsjö, 1982
Scotland (1981)	6.0 (3.2–11.3)	ND	1.9	+	Dennis, 1984
Baja, Mexico (1971–72)	1.7 (0.2–2.7)[a]	–1%	1.0	+	Henny & Anderson, 1979
Maryland, USA (1968–69)	3.0 (1.1–4.7)	–12%	1.0–1.3	+	Wiemeyer *et al.*, 1975
Florida, USA (1969–72)	1.2 (0.6–2.6)[a]	0%	0.8–1.0	+	Ogden, 1977
Idaho, USA (1972–73)	10.3 (2.0–20.1)	–14%[b]	1.3	+	Johnson *et al.*, 1975, Melquist, 1974
Massachusetts, USA (1980–84)	1.7 (0.2–2.3)	–7%	1.5–2.3	+	Poole, unpublished
Connecticut, USA (1983–84)	1.2 (0.3–2.3)	–16%	1.3	+	Poole, unpublished
Ontario, Canada (1967–71)	0.7 (0.1–1.0)	–1%	1.2	+	Grier *et al.*, 1977
New Jersey, USA (1971, 1974)	21.8 (7.5–67)	–16%	0.2–0.4	–	Wiemeyer *et al.*, 1978
Northeast USA (1969–71)	ca. 14 (3.1–55)[a]	–16%	0.5	–	Spitzer *et al.*, 1978
Corsica (1973)	ca. 30 (19–43)	ND	1.0	–	Terrasse & Terrasse, 1977

[a] DDE only.
[b] Fresh and addled eggs.

APPENDIX 8.

Levels of mercury (Hg) (ppb, dry weight) in Ospreys from Europe and the United States during the 1960s and 1970s.

A factor of 3.7 was used to convert figures reported as wet weight to dry weight (Häkkinen & Häsanen, 1980). Toxic effects are noted 'yes' only if egg, chick, or adult viability were jeopardized by mercury contamination.

Region (years)	Hg mean (range)	Analysis of	Toxic effects?	Study
Sweden (1971–73)	1040 (630–1680)	Eggs	Yes	Odsjö, 1982
Sweden (1974–82)	1030 (470–1750)	Eggs	No	Ahlgren & Eriksson, 1984
Finland (1972–80)	725 (300–1900)	Eggs	No	Häkkinen & Häsanen, 1980
Finland (1972–80)	2450 (400–6800)	Liver[a]	??	Häkkinen & Häsanen, 1980
Finland (1972–80)	2500 (100–8300)	Feathers	No	Häkkinen & Häsanen, 1980
Northeast USA (1969–76)	230 (?–?)	Eggs	No	Spitzer *et al.*, 1978
Eastern USA (1964–69)	440 (260–890)	Eggs	No	Wiemeyer *et al.*, 1975
Eastern USA (1964–73)	21 460 (2600–250 000)	Liver[b]	??	Wiemeyer *et al.*, 1980

[a] Nestlings
[b] Adults

APPENDIX 9.
How to build and site an Osprey nest platform.

For more details, write for the following report: Osprey Nest Platforms: Section 5.1.6, *US Army Corps of Engineers Wildlife Resources Management Manual* (Technical Report EL–86–21) (Martin *et al.*, 1986). This report can be ordered free of charge from: National Technical Information Service, 5285 Port Royal Rd, Springfield, VA 22161 USA; or from: Environmental Laboratory, US Army Engineer Waterways Experiment Station, PO Box 631, Vicksburg, MS 39180 USA.

Platforms

Nearly any type of platform can be used to support an Osprey nest. There are no strict design requirements. The following rough guidelines, however, are worth noting:

1. Size: about one meter (3.3 feet) square (top view) is adequate.
2. Materials: choose strong, durable wood, such as oak, exterior plywood, cypress, or cedar. Frame platforms should be made from lumber at least 2.5×7.5 cm (thickness × wide) ($= 1 \times 3$ inches). Make solid base nesting platforms from exterior plywood (or the equivalent) 1.9 cm (0.75 inch) thick.
3. Design: frame platforms work just as well as those with a solid base, as long as slats are not far apart. Oak shipping pallets, available free for the asking at most lumber yards, make ideal nest platforms with little or no modification needed.
4. Tying sticks or other nesting material to a platform before installing it helps to encourage Osprey settlement.

Supports

Figure A9.1 shows three different ways to support an Osprey nesting platform. Supports should be solid posts or trees, at least 10 cm (4 inches) in diameter. Height of nesting poles will vary depending on location (see below). A predator guard, smooth metal sheathing about 1.5 meters (4 ft 10 inches) long wrapped tightly around the pole and nailed in place just below the platform's braces, is a critical addition to any pole not sited over water. Metal cones (Figure A9.1, bottom) are probably more effective predator deterrents than sheathing, but they are also costlier and more difficult to install.

Tripods provide a simple, stable solution to supporting nest platforms over swampy ground or shallow water. Nesting pairs are strongly attracted to overwater nest sites, probably because such

Figure A9.1. Methods of mounting Osprey nesting platforms on poles and trees (*top*), or on tripods (*bottom*). Designs are from Martin *et al.* (1986) and T.U. Fraser, Sr (unpublished).

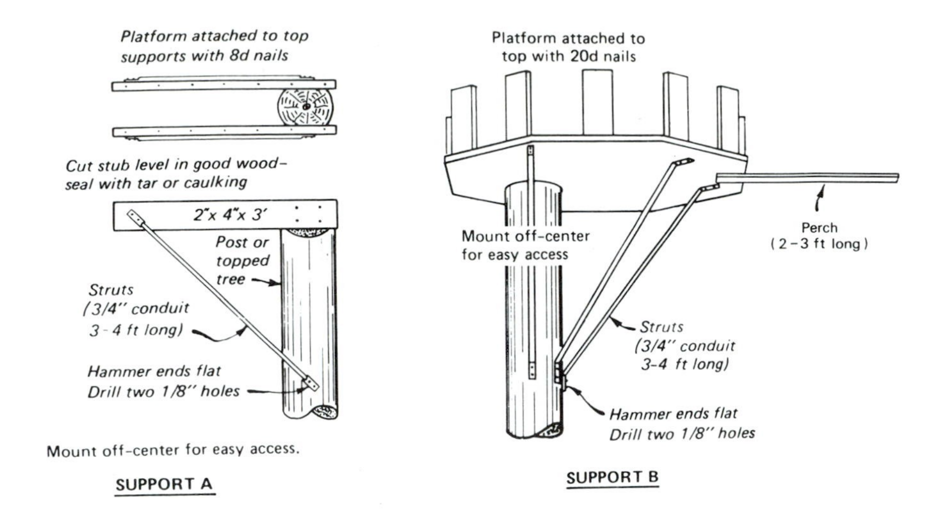

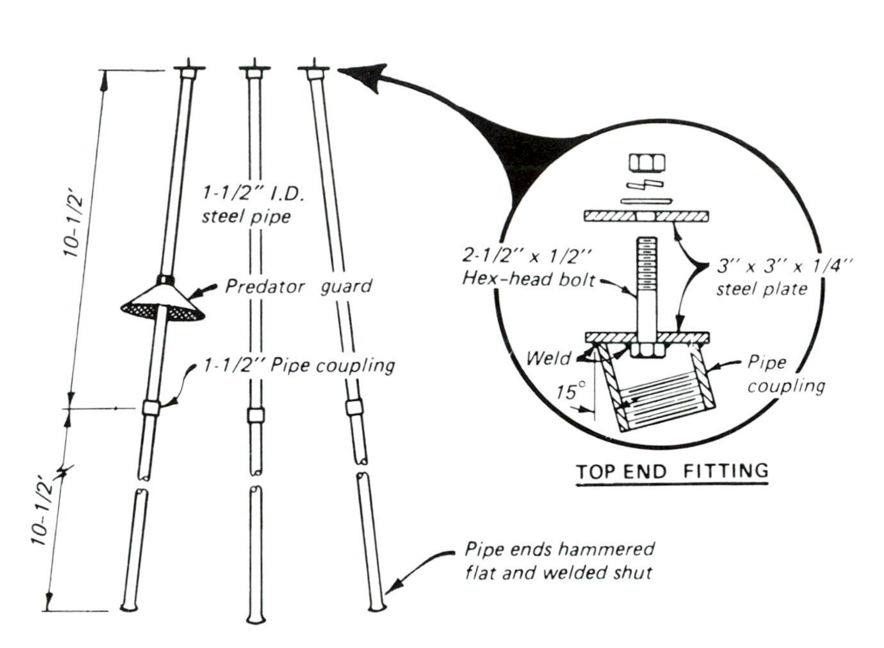

locations are safe from predators. See Webb & Lloyd (1984) for a variation on the tripod design shown above.

Siting a platform

Chapter 6 describes the natural sites where Ospreys choose to nest. Briefly, preferred natural sites are open areas (peninsulas, marshes, fields), overwater, or the tops of tall forest trees along shorelines. Sites near fish-rich waters have great appeal, particularly if other pairs nest nearby. Nesting poles in open locations need not be tall if they are well isolated from surrounding vegetation and well guarded from climbing predators (Fig. 10.3). In forest locations, platforms need more height (Fig. 10.1); those in the tallest trees are most likely to attract pairs. In all cases, it is best to locate platforms at least several hundred meters away from areas that people visit sporadically. Ospreys can habituate to nearby human activity, but do so only when exposed to continuous disturbance (Chapter 9). Begin by spacing platforms 200–300 meters apart, and an equal distance to active nests. As a colony grows, nests can be placed closer together.

It is difficult to overestimate the Osprey's attraction to nest sites surrounded by water. A platform secured to a solid piling or tripod 50–100 meters offshore makes an ideal nest site, although it must be well above storm tides and waves. Tripods are cheaper and easier to install than pilings but are more vulnerable to winter ice. Small, open, predator-free islands also attract pairs with great success. In short, isolate an Osprey nesting platform over water or on an island, and your chances of attracting a nesting pair will increase dramatically.

APPENDIX 10.
Major Osprey restoration projects completed or in progress before 1987, all in the United States.

A number of smaller projects were initiated during the mid–1980s in New Jersey, Kansas, West Virginia, and North Carolina, but results from these are not yet available.

Region	Dates	No. young hacked	No. young returned	Study
Northeast Pennsylvania	1980–1986	111	23	Rymon, unpublished; Schaadt & Rymon, 1983
Tennessee Valley	1979–1986	210	5–10?	Hammer & Beddow, 1984 and unpublished
Central Minnesota	1984–1986	35	1	Gillette & Englund, 1985 and unpublished
New York	1980–1986	30 +	2–3	B. Loucks, unpublished

References

Abbott, C.G. (1911). *The home life of the Osprey.* London: Witherby & Co.

Ahlgren, C-G. & Eriksson, M.O.G. (1984). Exposure to mercury and organochlorines of Osprey in southwest Sweden. *Vår Fågelvärld*, **43**, 299–305. (In Swedish, English summary.)

Airola, D.A. & Shubert, N. (1981). Reproductive success, nest site selection and management of Ospreys at Lake Almanor, California, 1969–1980. *Cal-Nevada Wildlife Transactions*, 79–89.

Ali, S. & Ripley, S.D. (1968). *Handbook of the birds of India and Pakistan*, vol. 1. London: Oxford University Press.

Allen, C.S. (1892). Breeding habits of the fish hawk on Plum Island, NY *Auk*, **9**, 313–21.

Almer, B., Dickson, W., Ekstrom, C., Hornstrom, E. & Miller, U. (1974). Effects of acidification on Swedish lakes. *Ambio*, **3**, 30–6.

Alt, K.L. (1980). Ecology of the breeding Bald Eagle and Osprey in the Grand Teton – Yellowstone National Parks complex. Unpublished MSc thesis, Montana State University, Bozeman.

Ames, P.L. (1966). DDT residues in the eggs of the Osprey in the northeastern USA and their relation to nest success. *Journal of Applied Ecology*, **3** (Supplement), 87–97.

Ames, P.L. & Mersereau, G.S. (1964). Some factors in the decline of the Osprey in Connecticut. *Auk*, **81**, 173–85.

Anonymous (1974). *Check list of Japanese birds.* Tokyo: Ornithological Society of Japan.

Austin-Smith, P.J. & Rhodenizer, G. (1983). Ospreys relocate nests from power poles to substitute sites. *Canadian Field Naturalist*, **97**, 315–19.

Bannerman, D.A. (1963). *Birds of the Atlantic Islands*, vol. 1. Edinburgh: Oliver and Boyd.

Bannerman, D.A. & Bannerman, W.M. (1968). *Birds of the Atlantic Islands*, vol. 4. Edinburgh: Oliver and Boyd.

Barradas, R.D. (1984). (Activities of a wintering population of Ospreys at the reservoir Miguel Aleman, Oaxaca, Mexico.) Unpublished MSc thesis, University of Veracruz. (In Spanish).

Barrowclough, G.F. (1983). Biochemical studies of microevolutionary processes. in *Perspectives in Ornithology*, ed. A.H. Brush & G.A. Clark, pp. 223–62. Cambridge: Cambridge University Press.

Beaman, M. & Galea, C. (1974). The visible migration of raptors over the Maltese Islands. *Ibis*, **116**, 419–31.

Beaman, M. & Porter, R.F. (1985). Status of birds of prey in Turkey. in *Bulletin of World Working Group on Birds of Prey*, No. 2. Berlin: ICBP.

Becker, J.J. (1985). *Pandion lovensis*, a new species of Osprey from the late Miocene of Florida. *Proceedings of the Biological Society of Washington*, **98**, 314–20.

Becsy, L. & Keve, A. (1977). The protection and status of birds of prey in Hungary. in *Proceedings of the World Conference on Birds of Prey*, ed. R.D. Chancellor, pp. 100–5. Vienna: ICBP.

Beer, C.G. (1980). The communication behavior of gulls and other seabirds. in *Behavior of marine animals* (IV): *marine birds*, ed. J. Burger, B. Olla, H. Winn. New York: Plenum Press.

Bent, A.C. (1937). *Life histories of North American Birds of prey: Order Falconiformes* (Part I). US National Museum Bulletin 167.

Berg, W., Johnels, A., Sjostrand, B. & Westermark, T. (1966). Mercury content in the feathers of Swedish birds from the past 100 years. *Oikos*, **17**, 71–83.

Berthon, D. & Berthon, S. (1984). (An account of the expedition 'Osprey' on the Mediterranean coast of Morocco.) *L'Oiseau et R.F.O.*, **54**, 201–13. (In French, English summary.)

Bider, J.R. & Bird, D.M. (1983). Distribution and densities of Osprey populations in the Great Whale region of Quebec. in *Biology and Management of Bald Eagles and Ospreys*, ed. D.M. Bird, pp. 223–30. Ste Anne de Bellevue, Quebec: Harpell Press.

Bijleveld, M. (1974). *Birds of prey in Europe.* London: Macmillan Press.

Bird, D.M. (chief ed.) (1983). *Biology and Management of Bald Eagles and Ospreys.* Ste Anne de Bellevue, Quebec: Harpell Press.

Blake, E.R. (1977). *Manual of Neotropical Birds*, vol. 1. Chicago: University of Chicago Press.

Blakers, M., Davies, S. & Reilly, P.N. (1984). *Atlas of Australian birds.* Melbourne: Melbourne University Press.

Blathwayt, F.L. (1921). Former breeding of the Osprey in Ireland. *British Birds*, **15**, 192.

Blus, L.J., Gish, C.D., Belisle, A.A. & Prouty, R.M. (1972). Logarithmic relationship of DDE residues to eggshell thinning. *Nature*, **235**, 376–7.

Bogan, J. & Newton, I. 1977. Redistribution of DDE in Sparrowhawks during starvation. *Bulletin of Environmental Contamination and Toxicology*, **18**, 317–22.

Boshoff, A.F. & Palmer, N.G. (1983). Aspects of the biology and ecology of the Osprey in the Cape Province, South Africa. *Ostrich*, **54**, 189–204.

Bouvet, F. & Thibault, J.-C. (1980). (Distribution, reproduction, and history of Ospreys in Corsica.) *Alauda*, **4**, 171–81. (In French, English summary.)

Bouvet, F. & Thibault, J.-C. (1981). (The status of the Osprey in Corsica.) *Rapaces méditerranées*, **6**, 104–7. (In French, English summary.)

Bradlee, T.S., Mowbray, L.L. & Eaton, W.F. (1931). Bermuda birds. *Proceedings of the Boston Society of Natural History*, **39**(8), 279–382.

Brodkorb, P. (1964). Catalogue of fossil birds, part 2 (Anseriformes through Galliformes). *Bulletin of the Florida State Museum*, **8**, 195–335.

Brown, L. (1970). *African birds of prey*. London: Collins.

Brown, L. (1980). *The African Fish Eagle*. Folkestone: Bailey Bros. & Swinfen, Ltd.

Brown, L. & Amadon, D. (1968). *Eagles, hawks, and falcons of the world*. vol. 2. London: Country Life Books.

Brown, L.R., Chandler, W., Flavin, C., Pollock, C., Postel, S., Starke, L. & Wolf, E.C. (1986). *State of the world, 1986*. New York: W.W. Norton & Co.

Brown, P.E. & Waterston, G. (1962). *The return of the Osprey*. London: Collins.

Bundy, G. & Warr, F.E. (1980). A check list of the birds of the Arabian Gulf States. *Sandgrouse*, **1**, 4–49.

Cade, T.J. (1982). *The Falcons of the World*. London: Comstock/Cornell University Press.

Carrier, W.D. & Melquist, W.E. (1976). The use of a rotor-winged aircraft in conducting nesting surveys of Ospreys in northern Idaho. *Raptor Research*, **10**, 77–83.

Cash, C.G. (1914). History of the Loch an Eilein Ospreys. *Scottish Naturalist*, **25**, 149–58.

Chapman, F.M. (1908). *Camps and cruises of an ornithologist*. New York: D. Appelton and Co.

Cheng, T. (1976). *Distribution list of Chinese birds*. Peking: Science Press.

Choate, E.A. (1985). *The dictionary of American bird names*, 2nd edn. Boston: Harvard Common Press.

Christensen, S., Lou, O., Muller, M. & Wohlmuth, H. (1981). The spring migration of raptors in southern Israel and Sinai. *Sandgrouse*, **3**, 1–42.

Clapham, H.S. (1964). The birds of the Dahlac Archipelago. *Ibis*, **106**, 376–88.

Clum, N.J. (1986). The effects of prey quality on the reproductive success of Osprey in the Adirondack mountains. Unpublished MSc thesis, Cornell University.

Collopy, M.W. (1984). Parental care, productivity, and predator-prey relationships of Ospreys in three north Florida lakes: preliminary report. in *Proceedings of the southeastern U.S. and Caribbean Osprey Symposium*, ed. M.A. Westall, pp. 85–98. Sanibel, FL: TIOF.

Conder, P. (1977). Legal status of birds of prey and owls in Europe. in *World Conference on Birds of Prey*, ed. R.D. Chancellor, pp. 189–92. Vienna: ICBP.

Cramp, S. & Simmons, K.E.L. (eds.) (1980). *The birds of the Western Palearctic*, vol. 2. Oxford: Oxford University Press.

Crosby, D.G. (1973). The fate of pesticides in the environment. *Annual Review of Plant Physiology*, **24**, 467–92.

Cupper, J. & Cupper, L. (1981). *Hawks in focus: a study of Australia's birds of prey*. Mildura, Australia: Jaclin Enterprises.

Curry-Lindahl, K. (1981). *Bird migration in Africa*, vol. 1. New York: Academic Press.

Dean, W.R.J. & Tarboton, W.R. (1983). Osprey breeding records in South Africa. *Ostrich*, **54**, 238–9.

Delacour, J. & Mayr, E. (1946). *Birds of the Philippines*. New York: Macmillan.

de Naurois, R. (1969). (Brief notes on the avifauna of the Cape Verde archipelago: fauna, endemism, ecology.) *Bull. Inst. Fond. Africa noire*, **31**, 143–218. (In French).

Dennis, R.H. (1983). Population studies and conservation of Ospreys in Scotland. in *Biology and Management of Bald Eagles and Ospreys*, ed. D.M. Bird, pp. 207–14. Ste Anne de Bellevue, Quebec: Harpell Press.

Dennis, R.H. (1984). *Birds of Badenoch and Strathspey*. Inverness: R. Dennis Enterprises.

DiCostanzo, J. (1980). Population dynamics of a Common Tern colony. *Journal of Field Ornithology*, **51**, 229–43.

Drent, R.H. (1975). Incubation. in *Avian Biology*, vol. 5, ed. D.S. Farner & J.R. King, pp. 333–420. New York: Academic Press.

Drent, R.H. & Daan, S. (1980). The prudent parent: energetic adjustments in avian breeding. *Ardea*, **68**, 225–52.

Dunstan, T.C. (1974). Feeding activities of Ospreys in Minnesota. *Wilson Bulletin*, **86**, 74–6.

duPont, J.E. (1971). *Philippine Birds*. Delaware Natural History Museum Monograph Series number 2.

Dyck, J., Eskildsen, J. & Moller, H.S. (1977). Status of birds of prey in Denmark 1975. in *Proceedings World Conference on Birds of Prey*, ed. R.D. Chancellor, pp. 30–3. Vienna: ICBP.

Eckstein, R.G. & Vanderschaegen, P.V. (1988). Ospreys in Wisconsin: distribution, reproductive success, and management. *Proceedings Eastern North American Osprey Symposium*. Sanibel, FL: TIOF.

Elowson, A.M. (1984). Spread-wing postures and the water repellency of feathers: a test of Rijke's hypothesis. *Auk*, **101**, 371–83.

Emlen, S.T. (1984). Cooperative breeding in birds and mammals. in *Behavioral Ecology: an evolutionary approach*. Sunderland, MA: Sinauer Associates.

Enderson, J.H., Craig, G.R., Burnham, W.A. & Berger, D.D. (1982). Eggshell thinning and organochlorine residues in Rocky Mountain Peregrines and their prey. *Canadian Field-Naturalist*, **96**, 255–64.

Eriksson, M.O.G. (1986). Fish delivery, production of young, and nest density of Osprey in southwest Sweden. *Canadian Journal of Zoology*, **64**, 1961–5.

Evans, P.R. & Lathbury, G.W. (1973). Raptor migration across the Straits of Gibraltar. *Ibis*, **115**, 572–85.

Fernandez, G. & Fernandez, J. (1977). Some instant benefits and long range hopes of color banding Ospreys. in *Transactions of the North American Osprey Research Conference*, ed. J. Ogden, pp. 89–94. US National Park Service.

Fischer, W. von. (1983). (Activities and daily schedules of Ospreys in southeast Asian winter quarters.) *Bietr. Vogelkd.*, **29**, 29–33. (In German.)

Fleming, J.W., Clark, D.R. & Henny, C.J. (1983). Organochlorine pesticides and PCBs: a continuing problem for the 1980s. in *Transactions of the North American Wildlife & Natural Resources Conference*, pp. 186–99. Washington, DC: Wildlife Management Institute.

Flook, D.R. & Forbes, L.S. (1983). Ospreys and water management at Creston, British Columbia. in *Biology and Management of Bald Eagles and Ospreys*, ed. D.M. Bird, pp. 281–6. Ste Anne de Bellevue, Quebec: Harpell Press.

Forbush, E.H. (1927). *Birds of Massachusetts and Other New England States*. Amherst: Mass Department of Agriculture.

Fuller, P.J. & Burbidge, A.A. (1981). *The birds of Pelsart Island, Western Australia*. Report No. 44, Department of Fisheries & Wildlife, Perth, Western Australia.

Gallagher, M.D. & Rogers, T.D. (1978). On the breeding birds of Bahrain. *Bonner Zoological Beiträge*, **29**, 5–17.

Gallagher, M.D. & Woodcock, W.M. (1980). *The Birds of Oman*. London: Quartet Books.

Galushin, V.M. (1977). Recent changes in the actual and legislative status of birds of prey in the USSR. in *Proceedings World Conference on Birds of Prey*, ed. R.D. Chancellor, pp. 152–9. Vienna: ICBP.

Galushin, V.M. & Flint, V.E. (eds.). (1983). (*Conservation of birds of prey*.) Moscow: Nauka. (In Russian.)

Garber, D.P., Koplin, J.R., & Kahl, J.R. (1974). Osprey management on the Lassen National Forest, California. in *Proceedings of the Conference on Raptor Conservation*, eds. F.N. Hammerstrom, Jr *et al.*, pp. 119–22. Raptor Research Report 2.

Gerrard, J.M., Whitfield, D.W. & Maher, W.J. (1976). Osprey-Bald Eagle relationships in Saskatchewan. *Blue Jay*, **34**, 240–6.

Gillette, L.N. & Englund, J.V. (1985). The Hennepin County Park Reserve District's Osprey Reintroduction Project. *Loon*, **57**, 52–8.

Glutz von Blotzheim, U.N., Bauer, K. & Bezzel, E. (1971). (*Handbook of the birds of Central Europe*, vol. 4.) Frankfurt: Akademische Verlagsgesellschaft. (In German.)

Gorban, I. (1985). Current data on the status of eagles in the western Ukraine, USSR. in *Bulletin of World Working Group on Birds of Prey*, no. 2. Berlin: ICBP.

Goulding, M. (1980). *The fishes and the forest: explorations in Amazonian natural history*. Berkeley: University of California Press.

Green, R. (1976). Breeding behaviour of Ospreys *Pandion haliaetus* in Scotland. *Ibis*, **118**, 475–90.

Greene, E.P. (1987). Information transfer at Osprey colonies: individuals discriminate between high and low quality information. *Nature*, **329**, 239–241.

Greene, E.P. & Freedman, B. (1986). Status of the Osprey in Halifax and Lunenburg Counties, Nova Scotia. *Canadian Field Naturalist*, **100**, 470–3.

Greene, E.P., Greene, A.E. & Freedman, B. (1983). Foraging behavior and prey selection by Ospreys in coastal habitats in Nova Scotia, Canada. in *Biology and Management of Bald Eagles and Ospreys*, ed. D.M. Bird, pp. 243–56. Ste Anne de Bellevue, Quebec: Harpell Press.

Greenwood, P.J. (1980). Mating systems, philopatry and dispersal in birds and mammals. *Animal Behaviour*, **28**, 1140–62.

Grier, J.W. (1982). Ban of DDT and subsequent recovery of reproduction in Bald Eagles. *Science*, **218**, 1232–5.

Grier, J.W., Sindlar, C.R. & Evans, P.L. (1977). Reproduction and toxicants in Lake of the Woods Ospreys. in *Transactions of the North American Osprey Research Conference*, ed. J. Ogden, pp. 181–92. US National Park Service.

Griffin, D.R. (1984). *Animal Thinking*. Cambridge, Massachusetts: Harvard University Press.

Grover, K.E. (1983). Ecology of the Osprey on the upper Missouri River, Montana. Unpublished MSc thesis, Montana State University, Bozeman.

Grover, K.E. (1984). Male-dominated incubation in Ospreys. *Condor*, **86**, 489.

Grubb, T.G. Jr (1977). Weather-dependent foraging in Ospreys. *Auk*, **94**, 146–9.

Gwinner, E. (1986). Internal rhythms in bird migration. *Scientific American*, **254**, 84–92.

Haga, A. (1981). (The Osprey in southeast Østfold: nest sites, population development, human traffic tolerance, and management.) *Fauna*, **34**, 101–9. (In Norwegian, English summary.)

Hagan, J.M. (1984). A North Carolina Osprey population: social group or breeding aggregation? in *Proceedings of the southeastern US and Caribbean Osprey Symposium*, ed. M.A. Westall, pp. 43–60. Sanibel, FL: TIOF.

Hagan, J.M. (1986). Colonial nesting in Ospreys. Unpublished PhD dissertation, North Carolina State University, Raleigh.

Häkkinen, I. (1978). Diet of the Osprey in Finland. *Ornis Scandinavica*, **9**, 111–16.

Häkkinen, I. & E. Häsanen. (1980). Mercury in the eggs and nestlings of the Osprey in Finland and its bioaccumulation from fish. *Annals Zoologica Fennica*, **17**, 131–9.

Hallberg, L.-D., Hallberg, P.-S. & Sondell, J. (1983). (Changing the location of Osprey nest sites to reduce human disturbance.) *Vår Fågelvärld*, **42**, 73–80. (In Swedish, English summary.)

Hammer, D.A. & Beddow, T.E. (1984). Hacking young Ospreys to restore Tennessee Valley populations. in *Proceedings of the southeastern U.S. and Caribbean Osprey Symposium*, ed. M.A. Westall, pp. 67–74. Sanibel, FL: TIOF.

Harrison, C.J.O. & Walker, C.A. (1976). Birds of the British Upper Eocene. *Zoölogical Journal of the Linnean Society*, **59**, 323–51.

Harvie-Brown, J.A. & Macpherson, H.A. (1904). *A vertebrate fauna of the North-west Highlands and Skye*. Edinburgh: David Douglas.

Hays, H. & Risebrough, R.W. (1972). Pollutant concentrations in abnormal young terns from Long Island Sound. *Auk*, **89**: 19–35.

Heintzelman, D.S. (1983). Variations in numbers of, and influence of intersecting diversion lines upon, Ospreys migrating along the Kittatinny Ridge in eastern Pennsylvania in autumn. *American Hawkwatcher*, **6**, 1–4.

Henny, C.J. (1977a). Research, management, and status of the Osprey in North America. in *Proceedings World Conference on Birds of Prey*, ed. R.D. Chancellor, pp. 199–222. Vienna: ICBP.

Henny, C.J. (1977b). California Ospreys begin incubation at a frozen mountain lake. *Bird-banding*, **48**, 274.

Henny, C.J. (1983). Distribution and abundance of nesting Ospreys in the USA. in *Biology and Management of Bald Eagles and Ospreys*, ed. D.M. Bird, pp. 175–86. Ste Anne de Bellevue, Quebec: Harpell Press.

Henny, C.J. & Anderson, D.W. (1979). Osprey distribution, abundance, and status in Western North America: III. The Baja California and Gulf of California population. *Bulletin of Southern California Academy of Sciences*, **78**(2), 89–106.

Henny, C.J., Byrd, M.A., Jacobs, J.A., McLain, P.D., Todd, M.R. & Halla, B.F. (1977). Mid-Atlantic coast Osprey population: present numbers, productivity, pollutant contamination, and status. *Journal of Wildlife Management*, **41**, 254–65.

Henny, C.J., Collins, J.A. & Diebert, W.J. (1978a). Osprey distribution, abundance, and status in Western North America: II. The Oregon population. *Murrelet*, **59**, 14–25.

Henny, C.J., Dunaway, D.J., Mulette, R.D. & Koplin, J.R. (1978b). Osprey distribution, abundance, and status in Western North America: I. The northern California population. *Northwest Science*, **52**, 261–71.

Henny, C.J. & Noltemeier, A.P. (1975). Osprey nesting populations in the coastal Carolinas. *American Birds*, **29**, 1073–9.

Henny, C.J. & Van Velzen, W.T. (1972). Migration patterns and wintering locations of American Ospreys. *Journal of Wildlife Management*, **36**, 1133–41.

Henny, C.J. & Wight, H.M. (1969). An endangered Osprey population: estimates of mortality and production. *Auk*, **86**, 188–98.

Hilton, J.R. (1977). Legal status of birds of prey in other parts of the world. in *Proceedings World Conference on Birds of Prey*, ed. R.D. Chancellor, pp. 193–5. Vienna: ICBP.

Hughes, J. (1986). *Distribution, abundance, and productivity of Ospreys in interior Alaska*. Juneau: Alaska Department of Fish & Game.

Hutchinson, G.E. (1978). *An introduction to population ecology*. New Haven: Yale University Press.

Immelmann, K. (1971). Ecological aspects of periodic reproduction. in *Avian Biology*, vol. 1, eds. D.S. Farner & J.R. King, pp. 341–89. New York: Academic Press.

Jacob, J.P., Jacob, A. & Courbet, B. (1980). (Spring observations of Osprey and Eleonora's Falcon along the Algerian coast.) *Le Gerfaut*, **70**, 405–8. (In French.)

Jamieson, I.G. & Seymour, N.R. (1983). Inter– and intra-specific agonistic behavior of Ospreys near their nest sites. *Canadian Journal of Zoology*, **61**, 2199–202.

Jamieson, I., Seymour, N.R. & Bancroft, R.P. (1982). Use of two habitats related to changes in prey availability in a population of Ospreys in northeastern Nova Scotia. *Wilson Bulletin*, **94**, 557–64.

Jensen, S., Johnels, A.G., Olsson, M. & Westermark, T. (1972). The avifauna of Sweden as indicators of environmental contamination with mercury and chlorinated hydrocarbons. *Proceedings of the International Ornithological Congress*, **15**, 455–65.

Johnson, D.R., Melquist, W.E. & Schroeder, G.J. (1975). DDT and PCB levels in Lake Coeur d'Alene, Idaho, Osprey eggs. *Bulletin of Environmental Contamination and Toxicology*, **13**, 401–5.

Judge, D.S. (1983). Productivity of Ospreys in the Gulf of California. *Wilson Bulletin*, **95**, 243–55.

Kahl, J.R. (1971). *Osprey habitat management plan, Lassen National Forest*. San Francisco: USDA Forest Service.

Kalaber, L. (1985). Status of diurnal birds of prey in Rumania and the problem of their protection. *Bulletin of the World Working Group on Birds of Prey*, no. 2. Berlin: ICBP.

Kerlinger, P. (1985). Water-crossing behavior of raptors during migration. *Wilson Bulletin*, **97**(1), 109–13.

Kerlinger, P., Cherry, J.D. & Powers, K.D. (1983). Records of migrant hawks from the North Atlantic Ocean. *Auk*, **100**, 488–90.

Kerlinger, P. & Gauthreaux, S.A. (1984). Flight behavior of Sharp-shinned Hawks during migration. I: over land. *Animal Behaviour*, **32**, 1021–8.

Ketterson, E.D. & Nolan, V., Jr (1983). The evolution of differential bird migration. in *Current Ornithology*, ed. R.F. Johnson, pp. 357–402. New York: Plenum Press.

King, J.R. (1974). Seasonal allocation of time and energy resources in birds. in *Avian Energetics*, ed. R.A. Paynter, pp. 4–70. Cambridge, Massachusetts: Nuttall Ornithological Club.

Knight, C.W.R. (1932). Photographing the nest life of the Osprey. *National Geographic*, **62**, 247–60.

Koivusaari, J., Laamanen, A., Nuuja, I., Palokangus, R. & Vihko, V. (1972). Notes on the concentration of some environmental chemicals

in the eggs of the White-tailed Eagle and Osprey in the Quarken area of the Gulf of Bothnia. *Work – Environment – Health*, **9**, 44–5.
Kurtén, B. (1980). *The dance of the tiger*. New York: Pantheon Books.
Kushlan, J.A. & Bass, O.L. (1983). Decreases in the southern Florida Osprey population, a possible result of food stress. in *Biology and Management of Bald Eagles and Ospreys*, ed. D.M. Bird, pp. 187–200. Ste Anne de Bellevue, Quebec: Harpell Press.
Lack, D. (1954). *The natural regulation of animal numbers*. Oxford: Oxford University Press.
Lack, D. (1968). Ecological adaptations for breeding in birds. London: Chapman & Hall.
Landsburg, H.E. (ed.) (1970, 1971, 1974). *World survey of climatology*, vols. 5, 11, 13. New York: American Elsevier.
Leopold, A. (1949). *Sand County Almanac*. Oxford: Oxford University Press.
Leshem, Y. (1984). Shell-dropping by Ospreys. *British Birds*, **78**, 143.
Levenson, H. (1979). Time and activity budgets of Ospreys nesting in northern California. *Condor*, **81**, 364–9.
Levenson, H. & Koplin, J.R. (1984). Effects of human activity on the productivity of nesting Ospreys. *Journal of Wildlife Management*, **48**, 1374–7.
Linear, M. (1982). Gift of poison – the unacceptable face of development aid. *Ambio*, **11**, 2–8.
Littrell, E.E. (1986). Shell thickness and organochlorine pesticides in Osprey eggs from Eagle Lake, California. *California Fish & Game*, **72**(3), 182–5.
Lobov, E.G. (1985). Raptor populations monitoring programme in Kamchatka. in *Bulletin of World Working Group on Birds of Prey*, no. 2, pp. 23–7. Berlin: ICBP.
Lofts, B. & Murton, R.K. (1973). Reproduction in birds. in *Avian Biology*, Vol. 3, ed. D. Farner & R.K. Murton. New York: Academic Press.
Love, J.A. (1983). *The Return of the Sea Eagle*. Cambridge: Cambridge University Press.
MacArthur, R.H. (1972). *Geographical Ecology*. New York: Harper & Row.
MacCarter, D.L. & MacCarter, D.S. (1979). Ten-year nesting status of ospreys at Flathead Lake, Montana. *Murrelet*, **60**, 42–9.
McLean, P.K. (1986). The feeding ecology of Chesapeake Bay Ospreys and the growth and behavior of their young. Unpublished MSc thesis, College of William and Mary, Williamsburg, Virginia.
Mackworth-Praed, C.W. & Grant, C.H.B. (1957). *Birds of eastern and northeastern Africa*, vol. 1. London: Longman, Green & Co.
Martin, C.O., Mitchell, W.A. & Hammer, D.A. (1986). Osprey nest platforms. Section 5.1.6 in *US Army Corps of Engineers Wildlife Resources Management Manual*, Technical Report EL-86–21. Vicksburg, Mississippi.
Mason, J. & Seip, H.M. (1985). The current state of knowledge on acidification of surface waters and guidelines for further research. *Ambio*, **14**, 45–51.

Medway, M.A. & Wells, D.R. (1976). *The birds of the Malay Peninsula*. London: Witherby.
Meinertzhagen, R. (1954). The education of young Ospreys. *Ibis*, **96**: 153–5.
Melotti, P. & Spagnesi, M. (1979). (Analysis of Osprey band recoveries in Italy, 1939–1977.) *Ricerche di Biologia della Selvaggina*, **65**, 1–19. (In Italian.)
Melquist, W.E. (1974). Nesting success and chemical contamination in northern Idaho and northwestern Washington Ospreys. Unpublished MSc thesis, University of Idaho, Boise.
Meyburg, B.-U. & Meyburg, C. (1987). (The Osprey as a breeding bird in Central Europe.) *(NF) Band*, **27**, 34–41. (In German.)
Mock, D.W. (1983). On the study of avian mating systems. in *Perspectives in ornithology*, ed. A.H. Brush & G.A. Clark, Cambridge: Cambridge University Press.
Moll, K.H. (1962). *Der fischadler*. (*The Osprey*.) Wittenberg, Lutherstadt: A. Ziemsen Verlag. (In German.)
Moreau, R.E. (1972). *The Palearctic-African bird migration systems*. New York: Academic Press.
Mowbray, D.L. (1986). Pesticide control in the southwest Pacific. *Ambio*, **15**, 22–9.
Mullen, P.D. (1985). Reproductive ecology of Ospreys in the Bitterroot Valley of Western Montana. Unpublished MSc thesis, University of Montana, Missoula.
Munro, G.C. (1960). *Birds of Hawaii*. Tokyo: Bridgeway Press.
Muntaner, J. (1981). (The status of Ospreys in the Balearic Islands). *Rapaces méditerranées*, pp. 100–3. (In French.)
NRC (National Research Council) (1978). *An assessment of mercury in the environment*. Environmental Studies Board, Commission on Natural Resources, National Academy of Sciences, Washington, DC.
Newton, I. (1979). *Population ecology of raptors*. Vermillion, South Dakota: Buteo Books.
Newton, I. & Marquiss, M. (1984). Seasonal trend in the breeding performance of Sparrowhawks. *Journal of Animal Ecology*, **53**, 809–30.
Nisbet, I.C.T. (1977). Courtship feeding and clutch size in Common Terns. in *Evolutionary Ecology*, ed. B. Stonehouse and C. Perrins, pp. 101–9. London: Macmillan.
Nisbet, I.C.T. (1980). Effects of toxic pollutants on productivity in colonial waterbirds. in *Transactions of the Linnaean Society of New York*, Vol. 9, pp. 103–14.
Nye, P.E. (1983). A biological and economic review of the hacking process for the restoration of Bald Eagles. in *Biology and Management of Bald Eagles and Ospreys*, ed. D.M. Bird, pp. 127–36. Ste Anne de Bellevue, Quebec: Harpell Press.
Nyholm, N.E.I. (1981). Evidence of involvement of aluminum in causation of defective formation of eggshells and of impaired breeding in wild passerine birds. *Environmental Research*, **26**, 363–71.
Odsjö, T. (1982). Eggshell thickness and levels of DDT, PCB, and mercury in eggs of Osprey and Marsh Harrier in relation to their

breeding success and population status in Sweden. Unpublished PhD dissertation, University of Stockholm.

Odsjö, T. & Sondell, J. (1976). Reproductive success of Ospreys in southern and central Sweden. *Ornis Scandinavica*, **7**, 71–84.

Odum, E.P. (1971). *Fundamentals of Ecology*, 3rd edition. Philadelphia: W.B. Saunders & Co.

Officer, C.B., Biggs, R.B., Taft, J.L., Cronin, E.L., Tyler, M.A. & Boynton, W.R. (1984). Chesapeake Bay anoxia: origin, development, and significance. *Science*, **223**, 22–7.

Ogden, J.C. (1975). Effects of Bald Eagle territoriality on nesting Ospreys. *Wilson Bulletin*, **87**, 496–505.

Ogden, J. (1977). Report on Florida Bay Ospreys. in *Transactions of the North American Osprey Research Conference*, ed. J. Ogden, pp. 143–51. US National Park Service.

Olendorff, R.R., Miller, A.D. & Lehman, R.N. (1981). Suggested practices for raptor protection on power lines – the state of the art in 1982. *Raptor Research Report*, no. 4. Raptor Research Foundation.

Österlöf, S. (1965). (Rapid decline of Ospreys in northern Iran.) *Die Vogelwarte*, **23**, 95–7. (In German.)

Österlöf, S. (1973). (The Osprey in Sweden, 1971.) *Vår Fågelvärld*, **32**: 100–106. (In Swedish, English summary.)

Österlöf, S. (1977). Migration, wintering areas, and site tenacity of the European Osprey (*Pandion h. haliaetus*). *Ornis Scandinavica*, **8**, 61–78.

Peakall, D.B. (1975). Physiological effects of chlorinated hydrocarbons on avian species. in *Environmental Dynamics of Pesticides*, ed. R. Haque & V.H. Freed, pp. 343–60. New York: Plenum Press.

Perfect, J. (1980). The environmental impact of DDT on a tropical agro-ecosystem. *Ambio*, **9**, 16–21.

Perrins, C.M. (1970). The timing of birds' breeding seasons. *Ibis*, **112**, 242–55.

Perrins, C.M. (1979). *British Tits*. London: Collins.

Perrins, C.M. & Birkhead, T.R. (1983). *Avian Ecology*. Glasgow: Blackie.

Peterson, R.T. (1969). The Osprey: endangered world citizen. *National Geographic*, **136**, 53–67.

Poole, A.F. (1979). Sibling aggression among nestling Ospreys in Florida Bay. *Auk*, **96**, 415–17.

Poole, A.F. (1981). The effects of human disturbance on Osprey reproductive success. *Colonial Waterbirds*, **4**, 20–7.

Poole, A.F. (1982a). Brood reduction in temperate and subtropical Ospreys. *Oecologia*, **53**, 111–19.

Poole, A.F. (1982b). Breeding Ospreys feed fledglings that are not their own. *Auk*, **99**, 781–4.

Poole, A.F. (1984). Reproductive limitation in coastal Ospreys: an ecological and evolutionary perspective. Unpublished PhD dissertation, Boston University.

Poole, A.F. (1985). Courtship feeding and Osprey reproduction. *Auk*, **102**, 479–92.

Poole, A.F. & Agler, B. (1987). Recoveries of Ospreys banded in the

United States, 1914–1984. *Journal of Wildlife Management*, **51**, 148–55.
Poole, A.F. & Shoukimas, J. (1982). A scale for weighing birds at habitual perches. *Journal of Field Ornithology*, **53**, 409–14.
Poole, A.F. & Spitzer, P.R. (1983). An Osprey revival. *Oceanus*, **26**, 49–54.
Postupalsky, S. (1974). Raptor reproductive success: some problems with methods, criteria, and terminology. *Raptor Research Report*, no. 2, 21–31.
Postupalsky, S. (1977). Status of the Osprey in Michigan. in *Transactions of the North American Osprey Research Conference*, ed. J. Ogden, pp. 153–67. US National Park Service.
Postupalsky, S. (1978). Artificial nesting platforms for Ospreys and Bald Eagles. in *Endangered birds: management techniques for preserving endangered species*, ed. S.A. Temple, pp. 35–45. Madison: University of Wisconsin Press.
Postupalsky, S. (1983). *1983 Bald Eagle and Osprey nesting surveys in Michigan*. Michigan Department of Natural Resources Report No. 2964.
Postupalsky, S. & Stackpole, S.M. (1974). Artificial nesting platforms for Ospreys in Michigan. in *Management of Raptors*, ed. F.N. Hammerstrom *et al.*, pp. 105–17. Raptor Research Foundation.
Prevost, Y.A. (1977). Feeding ecology of Ospreys in Antigonish County, Nova Scotia. Unpublished MSc thesis, McGill University.
Prevost, Y.A. (1982). The wintering ecology of Ospreys in Senegambia. Unpublished PhD dissertation, University of Edinburgh.
Prevost, Y.A. (1983a). Osprey distribution and subspecies taxonomy. in *Biology and Management of Bald Eagles and Ospreys*, ed. D.M. Bird, pp. 157–74. Ste Anne de Bellevue, Quebec: Harpell Press.
Prevost, Y.A. (1983b). The moult of the Osprey. *Ardea*, **71**, 199–209.
Prevost, Y.A., Bancroft, R.P. & Seymour, N.R. (1979). Status of the Osprey in Antigonish Co., Nova Scotia. *Canadian Field Naturalist*, **92**, 294–7.
Proctor, N.S. (1977). Osprey catches vole. *Wilson Bulletin*, **89**, 625.
Puleston, D. (1977). Osprey population studies on Gardiner's Island. in *Transactions of the North American Osprey Research Conference*, ed. J. Ogden, pp. 95–9. US National Park Service.
Rabanal, H.R. (1978). Forest conservation and aquaculture development of mangrove areas. in *Proceedings of the International Workshop on Mangrove and Estuarine Development*. pp. 145–52. PCARR, Los Banos, Laguna, Philippines.
Rand, A.L. & Gilliard, E.T. (1968). *Handbook of New Guinea Birds*. Garden City, NY: Natural History Press.
Ratcliffe, D.A. (1967). Decrease in eggshell weight in certain birds of prey. *Nature*, **215**, 208–10.
Rauscher, K.L. (1984). (An Osprey of the Middle Miocene from Austria.) *Beiträge zur Palaontologie Osterr.*, **11**, 61–9. (In German.)
Reese, J. (1969). A Maryland Osprey population 75 years ago and today. *Maryland Birdlife*, **25**, 116–19.

Reese, J. (1977). Reproductive success of Ospreys in Central Chesapeake Bay. *Auk*, **94**, 202–21.

Ricklefs, R.E. (1980). Geographical variation in clutch size among passerine birds: Ashmole's hypothesis. *Auk*, **97**, 38–49.

Risebrough, R.W. & Anderson, D. 1975. Some effects of DDE and PCB on mallards and their eggs. *Journal of Wildlife Management*, **39**, 508–13.

Rüppell, G. (1981). (Analysis of the prey-catching behavior of the Osprey.) *Journal für Ornithologie*, **122**, 285–305. (In German.)

Saïller, E. (1977). (The courtship and mating behavior of Ospreys in Corsica.) *Nos Oiseaux*, **34**, 65–72. (In French.)

Saïller, E. & Nardi, R. (1984). (Secret life of the Osprey.) *Airone*, **41**, 40–66. (In Italian.)

St John, C. (1849). *A tour in Sutherlandshire*. London: John Murray.

Sandeman, P.W. (1957). The rarer birds of prey; their status in the British Isles. *Pandion haliaetus*. British Birds, **50**, 129–55.

Saurola, P. (1975). *Persecution of raptors in Europe assessed by Finnish and Swedish ring recovery data*. ICBP Technical Publication No. 5. Berlin: ICBP.

Saurola, P. (1978). Artificial nest construction in Europe. in *Birds of prey management techniques*, ed. T.A. Geer, pp. 72–80. British Falconers Club.

Saurola, P. (1986). (The Osprey in Finland, 1971–1985.) *Lintumies*, **21**, 66–80. (In Finnish, English summary.)

Schaadt, C.P. & Rymon, L.M. (1982). Innate fishing behavior of Ospreys. *Raptor Research*, **16**, 61–2.

Schaadt, C.P. & Rymon, L.M. (1983). The restoration of Ospreys by hacking. in *Biology and Management of Bald Eagles and Ospreys*, ed. D. Bird, pp. 299–305. Ste Anne de Bellevue, Quebec: Harpell Press.

Schlatter, R.P. & Morales, J. (1980). (The Osprey in Chile, with special reference to Valdivia.) *Medio Ambiente* (Chile), **4**(2), 18–22. (In Spanish.)

Scott, F. & Houston, C.S. (1983). Osprey nesting success in west-central Saskatchewan. *Blue Jay*, **41**, 27–32.

Scott, F. & Surkan, D.L. (1976). An unsuspected Osprey concentration in west-central Saskatchewan. *Blue Jay*, **34**, 99–101.

Scott, M.L., Zimmerman, J.R., Marinsky, S., Mullenhoff, P.A., Rumsey, G.L. & Rice, R.W. (1975). Effects of PCBs, DDT, and mercury compounds on egg production, hatchability, and shell quality in Chickens and Japanese Quail. *Poultry Science*, **54**, 350–68.

Serventy, V. (1965). Osprey. *Pacific Discovery*, **18**, 11–15.

Sibley, C.G. & Ahlquist, J.E. (1972). A comparative study of the egg white proteins of non-passerine birds. *Peabody Museum Bulletin* (Yale University), **39**, 1–276.

Sibley, C.G. & Ahlquist, J.E. (1986). Reconstructing bird phylogeny by comparing DNA's. *Scientific American*, **254**, 82–94.

Simmons, R. (1986). Why is the foraging success of Ospreys wintering in southern Africa so low? *Gabar*, **1**, 14–19.

Smith, K.D. (1957). An annotated list of the birds of Eritrea. *Ibis*, **99**, 1–26.

Smythies, B.E. (1960). *The birds of Borneo*. Sabah: Sabah Society.
Spitzer, P.R. (1978). Osprey egg and nestling transfers: their value as ecological experiments and as management procedures. in *Endangered birds: management techniques for preserving threatened species*, ed. S.A. Temple, pp. 171–82. Madison: University of Wisconsin Press.
Spitzer, P.R. (1980). Dynamics of a discrete coastal breeding population of Ospreys in the northeastern USA, 1969–1979. Unpublished PhD thesis, Cornell University.
Spitzer, P.R., Poole, A.F. & Scheibel, M. (1983). Initial population recovery of breeding Ospreys in the region between New York City and Boston. in *Biology and Management of Bald Eagles and Ospreys*, ed. D.M. Bird, pp. 231–41. Ste Anne de Bellevue, Quebec: Harpell Press.
Spitzer, P.R., Risebrough, R.W., Grier, J.W. & Sindelar, C.R. (1977). Eggshell thickness – pollutant relationships among North American Ospreys. in *Transactions of the North American Osprey Research Conference*, ed. J. Ogden, pp. 13–20. US National Park Service.
Spitzer, P.R., Risebrough, R.W., Walker, W., Hernandez, R., Poole, A., Puleston, D. & Nisbet, I.C.T. (1978). Productivity of Ospreys in Connecticut–Long Island increases as DDE residues decline. *Science*, **202**, 333–5.
Springer, A.M., Walker II, W., Risebrough, R.W., Benfield, D., Ellis, D.H., Mattox, W.G., Mindell, D.P. & Roseneau, D.G. (1984). Origins of organochlorines accumulated by Peregrine Falcons, *Falco peregrinus*, breeding in Alaska and Greenland. *Canadian Field-Naturalist*, **98**, 159–66.
Sprunt, IV, A. (1977). Report on Osprey sightings and nest locations in coastal Mexico and British Honduras. in *Transactions of the North American Osprey Research Conference*, ed. J. Ogden, pp. 247–9. US National Park Service.
Stager, K.E. (1958). An Osprey in the mideastern Pacific Ocean. *Condor*, **60**, 257–8.
Stinson, C.H. (1977). Familial longevity in Ospreys. *Bird-Banding*, **48**, 72–3.
Stinson, C.H. (1978). The influence of environmental conditions on the time budgets of breeding Ospreys. *Oecologia*, **36**, 127–39.
Stinson, C.H. (1979). On the selective advantage of fratricide in raptors. *Evolution*, **33**, 1219–25.
Stocek, R.F. & Pearce, P.A. (1983). Distribution and reproductive success of Ospreys in New Brunswick, 1974–1980. in *Biology and Management of Bald Eagles and Ospreys*, ed. D.M. Bird, pp. 215–21. Ste Anne de Bellevue, Quebec: Harpell Press.
Stone, W. (1937). *Bird studies at old Cape May*. Philadelphia: Delaware Valley Ornithological Club.
Stone, W.B., Butkas, S.A. & Reilly, E.M. (1978). Long small intestine in the Osprey and Peregrine Falcon. *NY State Fish & Game Journal*, **25**, 178–81.
Stotts, V.D. & Henny, C.J. (1975). The age at first flight for young American Ospreys. *Wilson Bulletin*, **87**, 277–8.
Swenson, J.E. (1978). Prey and foraging behavior of Ospreys on

Yellowstone Lake, Wyoming. *Journal of Wildlife Management*, **42**(1), 87–90.

Swenson, J.E. (1979a). Factors affecting status and reproduction of Ospreys in Yellowstone National Park. *Journal of Wildlife Management*, **43**, 595–602.

Swenson, J.E. (1979b). The relationship between prey species ecology and dive success in Ospreys. *Auk*, **96**, 408–13.

Swenson, J.E. (1981). Status of the Osprey in southeastern Montana before and after the construction of reservoirs. *Western Birds*, **12**, 47–51.

Szaro, R.C. (1978). Reproductive success and foraging behavior of the Osprey at Seahorse Key, Florida. *Wilson Bulletin*, **90**, 112–18.

Taylor, P. (1986). Osprey captures Grey Squirrel. *Raptor Research*, **20**, 76.

Terrasse, J.F. & Terrasse, M. (1977). (The Osprey (*Pandion haliaetus*) in the western Mediterranean: distribution, census, reproduction, threats.) *Nos Oiseaux*, **34**, 111–27. (In French, English summary.)

Thorpe, C. & Boddam, A.D. (1977). Unusual diet for Osprey. *Ostrich*, **48**, 47.

Toft, C.A., Trauger, D.L. & Murdy, H.W. (1984). Seasonal decline in brood sizes of sympatric waterfowl and a proposed evolutionary explanation. *Journal of Animal Ecology*, **53**, 75–92.

Ueoka, M.L. & Koplin, J.R. (1973). Foraging behavior of Ospreys in northwestern California. *Raptor Research*, **7**(2), 32–8.

Valiela, I. (1984). *Marine ecological processes*. New York: Springer-Verlag.

VanDaele, L.J. & VanDaele, H.A. (1982). Factors affecting the productivity of Ospreys nesting in west-central Idaho. *Condor*, **84**, 292–9.

VanDaele, L.J., VanDaele, H.A. & Johnson, D.R. (1980). *Status and management of Ospreys nesting in Long Valley, Idaho*. Moscow, Idaho: University of Idaho. 49pp.

Vaurie, C. (1965). *The birds of the Palearctic Faunas: non-passeriformes*. London: Witherby.

Verner, J. & Willson, M.F. (1966). The influence of habitats on mating systems of North American passerine birds. *Ecology*, **47**: 143–7.

Ward, P. & Zahavi, A. (1973). The importance of certain assemblages of birds as 'information centers' for food finding. *Ibis*, **115**, 517–34.

Warter, S.L. (1976). A new Osprey from the Miocene of California (Falconiformes: Pandionidae). *Smithsonian Contributions to Paleobiology*, **27**, 133–9.

Wasser, J.S. (1986). The relationship of energetics of Falconiform birds to body mass and climate. *Condor*, **88**, 57–62.

Waterston, G. (1971). *Ospreys in Speyside*, 3rd edition. Edinburgh: RSPB.

Webb, W.L. & Lloyd, A.H. (1984). Design and use of tripods as Osprey nest platforms. in *Proceedings of the southeastern U.S. and Caribbean Osprey Symposium*, M.A. Westall, ed., pp. 99–108. Sanibel, FL: TIOF.

Westall, M.A. (ed.) (1984). *Proceedings of the southeastern U.S. and*

Caribbean Osprey Symposium. Sanibel, FL: TIOF. 132 pp.
Wetmore, A. (1965). *The birds of Panama*, part 1. Smithsonian Miscellaneous Collections, 150 (4617).
Wetmore, S.T. & Gillespie, D.I. (1976). Osprey and Bald Eagle populations in Labrador and northeastern Quebec, 1969–1973. *Canadian Field-Naturalist*, **90**, 330–7.
Wiemeyer, S.N. (1977). Reproductive success of Potomac River Ospreys, 1971. in *Transactions of the North American Osprey Research Conference*, ed. J. Ogden, pp. 115–20. US National Park Service.
Wiemeyer, S.N., Lamont, T.G. & Locke, L.N. (1980). Residues of environmental pollutants and necropsy data for eastern US Ospreys, 1964–1973. *Estuaries*, **3**, 155–67.
Wiemeyer, S.N., Spitzer, P.R., Krantz, W.C., Lamont, T.G. & Cromartie, E. (1975). Effects of environmental pollutants on Connecticut and Maryland Ospreys. *Journal of Wildlife Management*, **39**, 124–39.
Wiemeyer, S.N., Swineford, D.M., Spitzer, P.R. & McLain, P.D. (1978). Organochlorine residues in New Jersey Osprey eggs. *Bulletin of Environmental Contamination and Toxicology*, **19**, 56–63.
Wiley, J.W. (1984). Status of the Osprey in the West Indies. in *Proceedings of the southeastern U.S. and Caribbean Osprey Symposium*, ed. M. Westall, pp. 9–17. Sanibel, FL: TIOF.
Wiley, J.W. & Lohrer, F.E. (1973). Additional records of non-fish prey taken by Ospreys. *Wilson Bulletin*, **85**, 468–70.
Williams, T.C., Ireland, L.C. & Teal, J.M. (1977). Bird migration over the western North Atlantic Ocean. *American Birds*, **31**, 251–67.
Wilson, E.O. (1985). Time to revive systematics. *Science*, **230**, 1110.
Witt, H.-H., Juana, E., Varela, J. & Marti, R. (1983). (Ospreys in the Chaffarinas Islands (northeast coast of Morocco) – observations on breeding and feeding.) *Vogelwelt*, **104**, 168–75. (In German, English summary.)
Woodwell, G.M. (1967). Toxic substances and ecological cycles. *Scientific American*, **216**, 24–31.
Wotzkow, C. (1985). Status and distribution of Falconiformes in Cuba. in *Bulletin of the World Working Group on Birds of Prey*, no.2. Berlin: ICBP.
Zimmerman, D. (1984). Sanctuary on a powder keg. *International Wildlife* (Jan), 28–34.

Index